Boolean Algebra

Boolean Algebra

R. L. Goodstein

Dover Publications, Inc.
Minola, New York

Bibliographical Note

This Dover edition, first published in 2007, is an unabridged republication of the 1966 printing of the work originally published by Pergamon Press, London, in 1963.

International Standard Book Number: 0-486-45894-6.

Manufactured in the United States of America
Dover Publications, Inc., 31 East 2nd Street, Mineola, N.Y. 11501

Contents

Chapter I.   The Informal Algebra of Classes   1

Union, intersection and complementation. The
commutative, associative and distributive laws.
Difference and symmetric difference. Venn
diagrams.
Examples I   18

Chapter II.   The Self-Dual System of Axioms   21

Standard forms. Completeness of the axiom
systems. Independence of the axioms. An algebra
of pairs. Isomorphism and homomorphism.
Examples II   47

Chapter III.   Boolean Equations   49

A bi-operational system of axioms. Equivalence of
the two formalizations. The general solutions of
Boolean equations. Congruence relations.
Independence of the axioms.
Examples III   73

Chapter IV.   Sentence Logic   76

Sentence logic as a model of Boolean algebra. The
third system of axioms. Truth-table completeness.
Independence of the axioms. The deduction
theorem.
Examples IV   92

Chapter V.   **Lattices**   **94**

Partially ordered sets. Atoms, minimal and
maximal elements. Upper and lower bounds.
Distributive lattices. Complemented distributive
lattices. Union and intersection ideals. The lattice
of union ideals. Representation theorems for
Boolean algebras. A denumerable Boolean algebra.
Newman algebra.
Examples V   119

Solutions to Examples   122

Bibliography   137

Index   139

Preface

AFTER an informal introduction to the algebra of classes, three different axiomatizations are studied in some detail, and an outline of a fourth system of axioms is given in the examples. In the last chapter Boolean algebra is examined in the setting of the theory of partial order. The treatment is entirely elementary and my aim has been to use Boolean algebra as a simple medium for introducing important concepts of modern algebra. There is a large collection of examples, with full solutions at the end of the book.

I have used the symbols ∪, ∩ for union and intersection, but I have not introduced their current readings "cup" and "cap", which I find so unhelpful. I myself prefer to read "$A \cap B$" as "A and B", and "$A \cup B$" as "A or B" since the members of "$A \cup B$" are the member of A or of B, and the members of "$A \cap B$" are members of A and of B, and of course this reading is in conformity with the interpretation of Boolean algebra as an algebra of sentences.

I am happy to record my thanks to Mr. M. T. Partis for help in reading the proofs and to the compositors and printers for the excellence of their work.

R. L. GOODSTEIN.

University of Leicester.

CHAPTER ONE

The informal algebra of classes

1.0. Classes. Collections of objects, whether they are identified by a survey of their members or by means of some characteristic property which their members have, are called classes. The students in a particular room at a particular time form a class, the voters on an electoral roll of a certain town form a class (as do their names on the roll), the hairs on a man's head, the blood-cells in his body, the seconds of time he has lived, all these form classes. Featherless bipeds and mammals with the power of speech are classes characterized by common properties of their members; they are classes with a common membership, *equal* classes, as we shall say.

1.1. Membership. We shall use capital letters as names of classes. If an object a is a member of a class A we shall write

$$a \, \varepsilon \, A$$

and say that "a belongs to A", or "a is in A". The membership symbol "ε" (the Greek letter ε) is the initial letter of the Greek verb "to be". Thus "Earth ε Planets" expresses the relationship of our earth to the class of planets.

If a is not a member of a class A then we write

$$a \notin A.$$

If we can write down signs for all the members of a class we represent this class by enclosing the signs in brackets. Thus

{1, 2, 3} is the class containing the numbers 1, 2 and 3 (and nothing else), {2, 1, 3}, {3, 1, 2}, {1, 1, 2, 3} for instance denoting the same class, and {a, b, c, d} is the class containing just the first four letters of the alphabet. We can represent any fairly small class in this way, but the notation is obviously impractical for large classes (like the class of all numbers from 1 to 10^{10}) and meaningless for classes with an unlimited supply of members (like the class of all whole numbers).

The class whose sole member is some object A, namely the class $\{A\}$, must be distinguished from A itself. For instance if $A = \{1, 2\}$ then $\{A\}$ is a class with only one member, but A is a class with two members. A class with a single member is called a unit class. "The Master of Trinity" is a unit class, and so is "The Queen of England".

1.2. Inclusion. If every member of a class A is also a member of a class B we say that the class A is contained in the class B, or A is included in B, and write

$$A \subset B.$$

It is important to distinguish between the membership relation "ε" and the inclusion relation "$\subset$". The membership relation is the relation in which a member of a class stands to the class itself; on one side (the left) of the membership relation stands a class member, and on the other side (the right) stands a class. But inclusion is a relation between *classes*, and a class stands on each side of the relation of inclusion. If $A \subset B$, we say that A is a *subclass* of B, and that B is a *superclass* of A. Every class is included in itself, thus $A \subset A$, because the members of A (on the left) are necessarily members of the same class A (on the right). A subclass of a class A which is not just A itself, is called a *proper* subclass. If $A \subset B$ and $B \subset A$ then $A = B$, for every member of A is a member of B, and every member of B is a member of A, so that A and B have the same members.

1.3. The empty class and the universal class. A convenient fiction is the *empty*, or *null* class, the class without members. If no candidate presents himself for some examination, the class of candidates is the empty class. We denote the empty class by 0; thus the relation $x \, \varepsilon \, 0$ is false for every object x in the world. Another convenient fiction is the *universal* class, the class of everything (or everything under consideration) which we denote by 1. The null class and the universal class are each unique. The null class is considered to be a subclass of every class (for there is no object which is a member of 0 and not a member of any A). Any class is of course a subclass of the universal class. In particular $0 \subset 1$.

1.4. The complement of a class. If we remove from the universal class all the members of some class A, the objects which remain form the *class complement* of A, denoted by A'. The classes A, A' have no members in common, but everything in the universal class is either a member of A or a member of A'. The complement of the null class is the universal class, and conversely the complement of the universal class is the null class. That is

$$0' = 1, \qquad 1' = 0.$$

Complementation is *involutory*, that is to say the complement of the complement is the original class.

1.5. Union and intersection. Given two classes A, B we may form the class C, *called the union of A and B* whose members are precisely those objects which are members of A or members of B; if A and B have any members in common, these common members occur once only in the union. For instance if A and B are sacks of potatoes their union is formed by emptying both sacks into a third. The union of two classes A and B is denoted by

$$A \cup B.$$

By definition union is commutative, that is

$$A \cup B = B \cup A.$$

Examples

1. If $\qquad\qquad A = \{a, b, c, d\}$

and if $\qquad\qquad B = \{c, d, e, f\}$

then $\qquad\qquad A \cup B = \{a, b, c, d, e, f\}.$

2. If A is the class of even numbers and B is the class of odd numbers then $A \cup B$ is the class of all whole numbers.

3. If A is the class of cats and B the class of Persian cats then $A \cup B = A$, for every Persian cat is a cat.

4. If A is the class of cats and B is the class of cats with tails 5 ft long then $A \cup B = A$, for B is the null class and contributes nothing to the union.

For any class A,

$$A \cup 0 = A, \qquad A \cup 1 = 1, \qquad A \cup A = A.$$

For the members of $A \cup 0$ are *either* members of A, *or* members of 0, and 0 has no members. And the members of $A \cup 1$ include the members of 1, and so include everything.

Finally, the members of $A \cup A$ are just the members of A. The relation $A \cup A = A$ is called the *idempotent* law for union. Since every object belongs either to A or to A' it follows that

$$A \cup A' = 1.$$

The class of members common to two classes is called their *intersection*. The intersection of A, B is denoted by

$$A \cap B.$$

By definition, intersection is commutative, that is $A \cap B = B \cap A$.

Examples

1. If $A = \{a, b, c, d\}$, $B = \{c, d, e\}$

then $A \cap B = \{c, d\}$.

2. If A is the class of green-eyed cats, and B is the class of long-haired cats, then $A \cap B$ is the class of long-haired green-eyed cats.

3. If A is the class of cats and B the class of dogs then $A \cap B$ is the null class, for no creature is both cat and dog.

For any class A

$$A \cap 1 = A, \quad A \cap 0 = 0, \quad A \cap A = A.$$

For every member of A is common to A and the universal class, and the empty class has nothing in common with A (even if A itself is null). The third relation, the indempotent law for intersection, says just that every member of A is common to A and itself. Since A and A' have no member in common we have

$$A \cap A' = 0.$$

1.6. We proceed to establish some of the important relations which hold between complementation, inclusion, union and intersection.

1.61. We prove first that, for any classes A, B

$$A \cap B \subset A, \quad A \cap B \subset B$$

$$A \subset A \cup B, \quad B \subset A \cup B.$$

For the *common* members of A and B (if any) are members of A, and members of B, and the union $A \cup B$ consists of both the members of A and the members of B.

1.62. The three relations

(i) $A \subset B$,   (ii) $A \cup B = B$,   (iii) $A \cap B = A$,

are *equivalent*, that is to say, all three hold if any one of them holds. Let (i) hold:

then any member of $A \cup B$ is a member of B, or a member of A, and so of B, that is to say $A \cup B \subset B$, but $B \subset A \cup B$ and so (ii) holds; moreover every member of A is a common member of A, B so that $A \subset A \cap B$, and since $A \cap B \subset A$ therefore (iii) holds. Observe the technique by which we have proved an equation; to show that, say, $X = Y$, we prove both $X \subset Y$ and $Y \subset X$, or in words, every member of the left-hand class is a member of the right-hand class, and every member of the right-hand class is also a member of the left-hand class. Next let us suppose that (ii) holds:

since $A \subset A \cup B$ and $A \cup B = B$ therefore (i) holds, and hence (iii) holds. And if we are given (iii) then from $A \cap B \subset B$ follows (i) and hence (ii), which completes the proof.

1.63. De Morgan's laws. Union and intersection interchange under complementation.

More precisely,

$$(A \cup B)' = A' \cap B', \qquad (A \cap B)' = A' \cup B'.$$

These relations are called De Morgan's laws. It suffices to prove one of these relations, since each is an immediate consequence of the other, under complementation. We recall that the complement of the complement is the original set; from the first relation (with A', B' in place of A, B) we have

$$(A' \cup B')' = (A'' \cap B'')$$

that is

$$(A' \cup B')' = A \cap B$$

whence, taking the complements of both sides, (for if two classes are equal so are their complements)

$$(A' \cup B')'' = (A \cap B)',$$

that is

$$(A \cap B)' = A' \cup B'$$

as required.

We come now to the proof of the first relation.

If $c \, \varepsilon \, (A \cup B)'$ then $c \notin A \cup B$ and so $c \notin A$ and $c \notin B$, or in other words $c \, \varepsilon \, A'$ and $c \, \varepsilon \, B'$, so that $c \, \varepsilon \, A' \cap B'$, which proves that

(i) $$(A \cup B)' \subset A' \cap B'.$$

However, if $c \, \varepsilon \, A' \cap B'$ then $c \, \varepsilon \, A'$ and $c \, \varepsilon \, B'$, that is, $c \notin A$ and $c \notin B$ and therefore $c \notin A \cup B$, for all the members of the union are members of A or members of B. But if $c \notin A \cup B$ then $c \, \varepsilon \, (A \cup B)'$, which proves that

(ii) $$A' \cap B' \subset (A \cup B)'.$$

From the inclusions (i), (ii) we obtain the desired equality

$$(A \cup B)' = A' \cap B'.$$

1.7. The associative laws. Both union and intersection are associative, that is

$$A \cup (B \cup C) = (A \cup B) \cup C,$$

$$A \cap (B \cap C) = (A \cap B) \cap C.$$

To prove the associative law for union it suffices to observe that $A \cup (B \cup C)$ is the class of objects which belong to A or to B or to C, and $(A \cup B) \cup C$ is the same class. The associative law for intersection may be obtained from the associative law for union by means of De Morgan's laws, but it is simpler just to observe that $A \cap (B \cap C)$ is the class of members which are common to A, B and C, and this is the same class as $(A \cap B) \cap C$.

In virtue of the associative laws we may write $A \cup B \cup C$ for either of $(A \cup B) \cup C$, $A \cup (B \cup C)$ and $A \cap B \cap C$ for either of $(A \cap B) \cap C$, $A \cap (B \cap C)$. This freedom to omit brackets extends to any number of classes. For instance,

$$(A \cup B \cup C) \cup D = (A \cup B) \cup (C \cup D) = A \cup (B \cup C \cup D)$$

for each of these classes is the class whose members are the members of A, B, C, D, and no others, and so we may write $A \cup B \cup C \cup D$ for any of these classes. Since union (and intersection) are also commutative we may interchange the orde for classes in

$$A \cup B \cup C$$

at will. For instance

$$C \cup B \cup A = (C \cup B) \cup A = A \cup (C \cup B), \text{by the commutative law,}$$
$$= A \cup (B \cup C), \text{by the same law,}$$
$$= A \cup B \cup C.$$

This result clearly extends to any number of classes, for example

$$B \cup D \cup A \cup C = (B \cup D \cup A) \cup C$$
$$= (A \cup B \cup D) \cup C$$
$$= (A \cup B) \cup (D \cup C)$$
$$= (A \cup B) \cup (C \cup D)$$
$$= A \cup B \cup C \cup D,$$

and this is otherwise clear since both $B \cup D \cup A \cup C$ and $A \cup B \cup C \cup D$ are classes formed from the members of A, B, C and D (and no others).

1.8. The distributive laws.

Each of union and intersection is distributive over the other. Thus

$$A \cup (B \cap C) = (A \cup B) \cap (A \cup C)$$

and

$$A \cap (B \cup C) = (A \cap B) \cup (A \cap C).$$

These relations recall the distributive law of common arithmetic

$$a.(b + c) = a.b + a.c$$

but common arithmetic has only one distributive law [for $(a.b) + c$ is not generally equal to $(a + c).(b + c)$]. To prove that union is distributive over intersection we observe that if $x \varepsilon A \cup (B \cap C)$ then x belongs either to A or to $B \cap C$; if the former then

$x \varepsilon A \cup B$, and $x \varepsilon A \cup C$ and so $x \varepsilon (A \cup B) \cap (A \cup C)$; if the latter then $x \varepsilon B$ and $x \varepsilon C$ and so $x \varepsilon A \cup B$ and $x \varepsilon A \cup C$ and again $x \varepsilon (A \cup B) \cap (A \cup C)$, which proves that

(i) $$A \cup (B \cap C) \subset (A \cup B) \cap (A \cup C).$$

Conversely, if $x \varepsilon (A \cup B) \cap (A \cup C)$ then $x \varepsilon A \cup B$ and $x \varepsilon A \cup C$; if $x \notin A$ then necessarily $x \varepsilon B$ and $x \varepsilon C$ so that $x \varepsilon B \cap C$ and finally $x \varepsilon A \cup (B \cap C)$, and if $x \varepsilon A$ then it remains true that $x \varepsilon A \cup (B \cap C)$ which proves that

(ii) $$(A \cup B) \cap (A \cup C) \subset A \cup (B \cap C)$$

and from (i), (ii) the first distributive law follows.

The second distributive law may be proved in the same way or may be derived from the first by complementation.

1.81. The twin relations

$$A \cup B = 1, \qquad A \cap B = 0$$

hold if, and only if, $B = A'$. We have already remarked that $A \cup A' = 1$, $A \cap A' = 0$, and so it remains to prove that A' alone has this property.

From $A \cup B = 1$ follows

$$A' \cap (A \cup B) = A' \cap 1 = A'$$

and so, by the distributive law

$$A' = (A' \cap A) \cup (A' \cap B)$$
$$= 0 \cup (A' \cap B) = A' \cap B$$

whence, by 1.62, $A' \subset B$.

From $A \cap B = 0$ we obtain

$$A' \cup (A \cap B) = A' \cup 0 = A'$$

and so

$$(A' \cup A) \cap (A' \cup B) = A'$$

whence

$$1 \cap (A' \cup B) = A'$$

and so $$A' \cup B = A'$$

and now from 1.62 it follows that $B \subset A'$.
Thus from both the relations

$$A \cup B = 1, \qquad A \cap B = 0$$

we derive both

$$B \subset A', \qquad A' \subset B$$

that is,

$$B = A'.$$

1.82. For any classes A, B, C, (i) if $A \subset B$ and $A \subset C$ then
$A \subset B \cap C$, and (ii) if $A \subset C$, $B \subset C$ then $A \cup B \subset C$.
For if (i) $A \subset B$ and $A \subset C$ then $A \cap B = A$, $A \cap C = A$ and so

$$A \cap (B \cap C) = (A \cap B) \cap C = A \cap C = A$$

so that $A \subset B \cap C$; and if (ii) $A \subset C$ and $B \subset C$ then $A \cup C = C$,
$B \cup C = C$ and so $(A \cup B) \cup C = A \cup (B \cup C) = A \cup C = C$
proving that $A \cup B \subset C$.
1.83. If $A \subset B$ then

$$A \cap C \subset B \cap C,$$

and

$$A \cup C \subset B \cup C.$$

For $A = A \cap B$ and so $A \cap C = A \cap B \cap C$ whence

$$(A \cap C) \cap (B \cap C) = A \cap (B \cap C) \cap (B \cap C)$$
$$= A \cap B \cap C = A \cap C$$

which proves that $\qquad A \cap C \subset B \cap C$;

and since $\qquad A \cup B = B, \qquad$ we have

$$(A \cup C) \cup (B \cup C) = (A \cup B) \cup (C \cup C) = (A \cup B) \cup C = B \cup C$$

proving that $\qquad A \cup C \subset B \cup C.$

A consequence of these results is that if $A = B$ then $A \cap C = B \cap C$, and $A \cup C = B \cup C$. For if $A = B$ then $A \subset B$ and $B \subset A$.

1.84. It follows from 1.61 and 1.83 that

$$A \cap (A \cup B) = A,$$

for $A \cap (A \cup B) \subset A$, and $A \subset A \cup B$ by 1.61,

so that $\qquad A = A \cap A \subset A \cap (A \cup B)$.

1.85. A necessary and sufficient condition for $A \subset B$ is that $A \cap B' = 0$; for if $A \subset B$ then $A = A \cap B$ and so $A \cap B' = A \cap (B \cap B') = A \cap 0 = 0$, and conversely if $A \cap B' = 0$ then

$$A = A \cap 1 = A \cap (B \cup B') = (A \cap B) \cup (A \cap B')$$
$$= (A \cap B) \cup 0 = A \cap B.$$

1.86. If for all classes A, $B \subset A$ then $B = 0$; for in particular $B \subset 0$; but $0 \subset B$ and so $B = 0$.
If for all classes A, $A \subset B$ then $B = 1$, for in particular $1 \subset B$; but $B \subset 1$ and so $B = 1$.

1.87. If $\qquad\qquad A \cup B = 0$

then $A = 0$ and $B = 0$; for $A \subset A \cup B = 0$ so that
$A = \qquad\qquad 0$, and similarly $B = 0$.

1.88. If $\qquad\qquad A \cap B = 1$

then $A = 1$ and $B = 1$; for $1 = A \cap B \subset A$ so that $A = 1$, and similarly $B = 1$.

1.9. The *difference* $A - B$ between two classes A, B is defined as the class of all elements of A which are not elements of B, that is

$$A - B = A \cap B'.$$

We proceed to examine some properties of class-difference. We observe first that

$$1 - A = A',$$

for $1 - A = 1 \cap A' = A'$; and in particular $1 - 0 = 0' = 1$.

1.91. The two relations

$$A - B = 0, \qquad A \subset B$$

are equivalent, for if $A \subset B$ then $A \cap B' = 0$, that is $A - B = 0$, by 1.85, and conversely.

1.92. An important relationship between union and difference is

$$(A - B) \cup B = A \cup B;$$

for $(A - B) \cup B = (A \cap B') \cup B = (A \cup B) \cap (B' \cup B)$
$$= (A \cup B) \cap 1 = A \cup B.$$

In particular, if $B \subset A$ then

$$(A - B) \cup B = A$$

for if $B \subset A$ then $A \cup B = A.$

Moreover, $A - B = A$ if and only if $A \cap B = 0$. For if $A - B = A$ then $A = A \cap B'$ and so

$$A \cap B = A \cap B' \cap B$$
$$= A \cap (B' \cap B)$$
$$= A \cap 0 = 0,$$

and if $A \cap B = 0$ then

$$A = A \cap 1 = A \cap (B \cup B')$$
$$= (A \cap B) \cup (A \cap B') = 0 \cup (A \cap B')$$
$$= A \cap B' = A - B.$$

1.93. Intersection is distributive over difference, that is

$$C \cap (A - B) = (C \cap A) - (C \cap B);$$
for $(C \cap A) - (C \cap B) = (C \cap A) \cap (C \cap B)'$
$$= (C \cap A) \cap (C' \cup B')$$
$$= (C \cap A \cap C') \cup (C \cap A \cap B')$$
$$= C \cap A \cap B', \text{ since } C \cap C' = 0,$$
$$= C \cap (A - B).$$

It is not however true that union is distributive over difference; this is clear from the union considered in 1.92 because the class $(A - B) \cup B$ contains all the elements of B whereas the class

$(A \cup B) - (B \cup B) = (A \cup B) \cap B' = (A \cap B') \cup (B \cap B') = A \cap B'$, contains no element of B.

1.94. From the difference between two classes $A - B$, we construct the *symmetric difference*

$$A + B$$

which is defined as the union of the difference between A and B, and the difference between B and A, that is

$$A + B = (A - B) \cup (B - A) = (A \cap B') \cup (B \cap A').$$

The symmetric difference may be expressed in various ways. For instance

$$A + B = (A \cup B) \cap (A' \cup B').$$

For

$$
\begin{aligned}
(A \cup B) \cap (A' \cup B') &= \{(A \cup B) \cap A'\} \cup \{(A \cup B) \cap B'\} \\
&= (A \cap A') \cup (B \cap A') \cup (A \cap B') \cup (B \cap B') \\
&= (B \cap A') \cup (A \cap B') = A + B.
\end{aligned}
$$

Another representation is

$$A + B = (A \cup B) - (A \cap B)$$

which shows clearly the nature of the symmetric difference as the class of elements which belong to A or to B but not to both, and motivates the choice of the addition sign for the symmetric difference.

To prove this relation we observe that

$$
\begin{aligned}
(A \cup B) - (A \cap B) &= (A \cup B) \cap (A \cap B)' \\
&= (A \cup B) \cap (A' \cup B') = A + B.
\end{aligned}
$$

Another important relation which follows immediately from the definition is

$$A' + B' = A + B.$$

1.95. Symmetric difference is of course commutative, as its name implies, for

$$B + A = (B - A) \cup (A - B) = (A - B) \cup (B - A) = A + B.$$

It is also associative, that is

$$(A + B) + C = A + (B + C).$$

Since $$A + B = (A \cup B) \cap (A' \cup B')$$

therefore

$$(A + B)' = (A \cap B) \cup (A' \cap B')$$

and so

$$
\begin{aligned}
(A + B) + C &= \{(A+B)\cup C\} \cap \{(A+B)'\cup C'\} \\
&= (A\cup B\cup C)\cap(A'\cup B'\cup C)\cap\{C'\cup(A\cap B)\cup(A'\cap B')\} \\
&= (A\cup B\cup C)\cap(A'\cup B'\cup C)\cap\{C'\cup A\cup(A'\cap B')\}\cap \\
&\qquad\qquad\qquad\qquad\qquad\qquad \{C'\cup B\cup(A'\cap B')\} \\
&= (A\cup B\cup C)\cap(A'\cup B'\cup C)\cap(A\cup B'\cup C')\cap(A'\cup B\cup C')
\end{aligned}
$$

and this expression is *unchanged* by interchanging A, C (the first bracket becomes $C \cup B \cup A$ which is equal to $A \cup B \cup C$, the second and third brackets exchange contents and the fourth bracket becomes $C' \cup B \cup A'$ which again equals $A' \cup B \cup C'$) so that

$$(A + B) + C = (C + B) + A;$$

but $C + B = B + C$ and $(B + C) + A = A + (B + C)$ and so

$$(A + B) + C = A + (B + C).$$

1.96. Intersection distributes our symmetric difference; i.e.

$$C \cap (A + B) = (C \cap A) + (C \cap B).$$

For
$$
\begin{aligned}
C \cap (A + B) &= C \cap \{(A - B) \cup (B - A)\} \\
&= \{C \cap (A - B)\} \cup \{C \cap (B - A)\} \\
&= \{(C \cap A) - (C \cap B)\} \cup \{(C \cap B) - (C \cap A)\} \\
&= (C \cap A) + (C \cap B).
\end{aligned}
$$

It is not, however, true that union distributes over symmetric difference; for instance,

$$A \cup (A + B) = A \cup B$$

but

$$(A \cup A) + (A \cup B) = A + (A \cup B) = B \cap A'.$$

1.97. $A + A = 0$

For $A + A = (A \cap A') \cup (A' \cap A) = 0 \cup 0 = 0$.

1.971. If $A + B = 0$ then $B = A$; for $A + B = (A \cap B') \cup (A' \cap B)$, and so from $A + B = 0$ we deduce $A \cap B' = 0$, $A' \cap B = 0$ so that $A \subset B$ and $B \subset A$ and therefore $A = B$.

1.972. $A_i = B_i$, $i = 1, 2, \dots, n$ if and only if

$$(A_1 + B_1) \cup (A_2 + B_2) \cup \dots \cup (A_n + B_n)' = 0;$$

for each equation $A_i = B_i$ is equivalent to $A_i +^? B_i = 0$.

1.98. We define the *cross* of A and B, denoted by $A \times B$, to be the complement of $A + B$, so that

$$A \times B = (A \cap B) \cup (A' \cap B').$$

It follows immediately that

$$A' \times B' = A \times B$$

and from 1.95, by taking complements, that

$$A \times B = B \times A$$

$$(A \times B) \times C = A \times (B \times C).$$

From 1.96 we conclude that union distributes over cross, and from 1.97 that

$$A \times A = 1.$$

We observe that, since

$$(A + B)' = (A' \cup B) \cap (B' \cup A)$$

therefore

$$A \times B = A' + B = A + B'$$

whence we conclude

$$A \times B' = A' \times B = A + B.$$

Moreover

$$A \times 1 = A + 0 = (A \cup 0) \cap (A' \cup 1) = A \cap 1 = A$$

and

$$A \times 0 = A + 1 = (A \cup 1) \cap (A' \cup 0) = 1 \cap A' = A'.$$

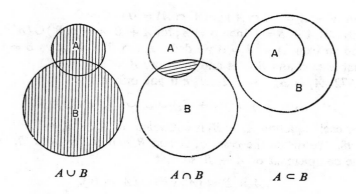

$A \cup B$ $A \cap B$ $A \subset B$

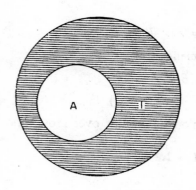

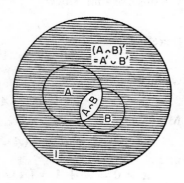

Finally we conclude from 1.971 that

$$A \times B = 1$$

if and only if

$$B = A.$$

1.99. Venn diagrams. The representation of classes by overlapping circles, known as Venn diagrams (or Euler diagrams) is a valuable visual aid in the understanding of class relations. The figures opposite illustrate the use of these diagrams to represent union, intersection and inclusion (*see top illustration opposite*). If we denote the universal class also by a circle, then the complement of a class is the part of circle 1 outside circle A (*see central illustration opposite*—the shaded region depicts the complement of A). In the *bottom illustration opposite*, the shaded region is both the part of 1 exterior to $A \cap B$ and the union of the exteriors of A and B.

In a similar way we may represent the difference and symmetric difference of two classes.

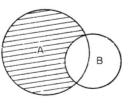
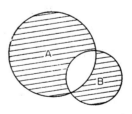

FIG. 1   FIG. 2

In Fig. 1 the shaded region represents $A - B$.
In Fig. 2 the shaded region represents $A + B$.

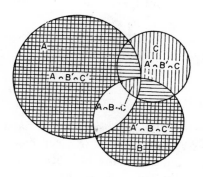

In the above diagram the region with horizontal shading is $A + B$ and that with vertical shading is $(A + B) + C$. The region shaded both horizontally and vertically is

$$(A \cap B' \cap C') \cup (A' \cap B \cap C').$$

The diagram reveals that
$$(A + B) + C =$$
$$(A \cap B' \cap C') \cup (A' \cap B \cap C') \cup (A' \cap B' \cap C) \cup (A \cap B \cap C).$$

EXAMPLES I

1. Prove that $A \subset B$ is equivalent to $A' \cup B = 1$.
2. Prove that $(A \cup B \cup C)' = A' \cap B' \cap C'$, $(A \cap B \cap C)' = A' \cup B' \cup C'$ and generalize to n classes.
3. If $X \subset A$ and $X \subset A'$ prove that $X = 0$;
   if $A \subset X$ and $A' \subset X$ prove that $X = 1$;
   if $A \subset B$, $C \subset D$ prove $A \cup C \subset B \cup D$.
4. If it is false that $A \subset B'$ prove that there is an $X \neq 0$, such that $X \subset A$ and $X \subset B$.
5. Prove that $A + B = A' + B'$.
6. If $A + K = B + K$, prove that $A = B$.
   If $A + B = 0$, prove that $B = A$.
7. Prove that $(A + B) + (C + D) = (A + C) + (B + D)$.
8. Prove that $(A + B)' = A' + B = A + B'$
$$(A - K) \cup (B - K) = (A \cup B) - K$$
and $\qquad (A + K) \cup (B + K) = (A \cap B) + K \cup (A + B)$.

9. Show that
$$A + (A \cup B) = B + (A \cap B) = B - (A \cap B)$$
$$A \cup B = A + B + A \cap B$$
and illustrate these relations by Venn diagrams.

10. If $A \cap B = 0$ prove that $A + B = A \cup B$.

11. Prove that
$$(A + B) \times C = (A \times C) + B = A + (B \times C)$$
and
$$(A \times C) + (B \times C) = A + B.$$

12. Prove the following relations
.1 $\qquad (A - B) + B = A \cup B$
.2 $\qquad (A - B) \cap B = 0$
.3 $\qquad A \cap (A - B) = A - B$
.4 $\qquad A - B \subset A$
.5 $\qquad A - A = 0$
.6 $\qquad A - (B - C) = (A - B) \cup (A \cap C)$
.7 $\qquad A - (A - B) = A \cap B$
.8 $\qquad (A - B) - C = (A - C) - (B - C)$
.9 $\qquad A - (B \cap C) = (A - B) \cup (A - C)$
.91 $\qquad A - (B \cup C) = (A - B) \cap (A - C)$
.92 $\qquad A - B = (A \cup B) - B = A - (A \cap B)$
.93 $\qquad (A \cup B) \cup (B - A) = A \cup B$

13. Prove the equivalence of the following pairs of relations
.1 $\qquad A \cup B = 0, \qquad A = 0$ and $B = 0$;
.2 $\qquad A - B = A, \qquad B - A = B$;
.3 $\qquad A \cup B = A - B, \qquad B = 0$;
.4 $\qquad A - B = A \cap B, \qquad A = 0$;
.5 $\qquad A \cup B \subset C, \qquad A \subset C$ and $B \subset C$;
.6 $\qquad C \subset A \cap B, \qquad C \subset A$ and $C \subset B$;
.7 $\qquad A \subset B \cup C, \qquad A - B \subset C$;
.8 $\qquad A - B = B - A, \qquad A = B$;
.9 $\qquad A \cap B = A \cup B, \qquad A = B$;
.91 $\qquad A \subset B \subset C, \qquad A \cup B = B \cap C$;
.92 $A \subset B$ and $C \subset D, \qquad (A - B) \cup (C - D) = 0$;
.93 $\qquad A = B, \qquad (A - B) \cup (B - A) = 0,$
$\quad (A = B)$ and $(C = D), \qquad (A-B) \cup (B-A) \cup (C-D) \cup (D-C) = 0$;
.94 $\qquad A - X = B - X, \qquad (A + B) \subset X.$

14. Prove that
.1 $(A \cup B) \cap (B \cup C) \cap (C \cup A) = (A \cap B) \cup (B \cap C) \cup (C \cap A)$
.2 $(A \cup B \cup C) \cap (B \cup C \cup D) \cap (C \cup D \cup A) \cap (D \cup A \cup B) =$
$\quad = (A \cap B) \cup (A \cap C) \cup (A \cap D) \cup (B \cap C) \cup (B \cap D) \cup (C \cap D)$
.3 $\qquad A - (B \cup C) = (A - B) - C$
.4 $\qquad (A - B) \cap C = (A \cap C) - B$
.5 $\qquad (A \cup B) - C = (A - C) \cup (B - C)$

.6 $A - (B - A) = A$

.7 $(A - C) \cap (B - C) = (A \cap B) - C$

.8 $(A - B) \cap (C - D) = (A \cap C) - (B \cup D)$

.9 $A - [B - (C - D)] = (A - B) \cup [(A \cap C) - D]$

.91 $A \cup B \cup C = (A - B) \cup (B - C) \cup (C - A) \cup (A \cap B \cap C)$

15. Find a set of six classes which includes A and B and is such that the difference of any two classes in the set is also in the set.

16. Prove that

$$(A \cup C) \subset (A \cup B) \cup (B \cup C)$$
$$A - C \subset (A - B) \cup (B - C)$$
$$A + C \subset (A + B) \cup (B + C)$$

17. Show that

$$A - D \subset (A - B) \cup (B - C) \cup (C - D).$$

CHAPTER TWO

The self-dual system of axioms

2. All the properties of classes which we established in the previous chapter may be derived from a few basic properties, *without making any use whatever of the class concept.*

As basic properties we take first the commutative and distributive laws for each of two operations $\cup$, $\cap$ (corresponding to union and intersection).

2.01. $A \cup B = B \cup A, \quad A \cap B = B \cap A$

2.02. $A \cup (B \cap C) = (A \cup B) \cap (A \cup C),$

$A \cap (B \cup C) = (A \cap B) \cup (A \cap C).$

Next we suppose that there are two distinct "classes" 0, 1 such that

2.03. $A \cup 0 = A, \quad A \cap 1 = A$

for any A, and that to each class A corresponds a class A' such that

2.04. $A \cup A' = 1, \quad A \cap A' = 0.$

We may continue to talk of classes 0, 1, A, B, C, ... and of union, intersection and complementation, but we shall not make use of any properties of these notions other than those listed in 2.01–2.04.

2.1. **Equality.** We suppose, further, that if $A = B$ then $A' = B'$, $A \cup C = B \cup C$ and $A \cap C = B \cap C$ for any class C, and if $A = B$ and $B = C$ then $B = A$ and $A = C$.

2.2. We start by showing that 0, 1 are each unique and that the complement of a class is unique. To prove that the zero is unique we suppose that *two* elements 0 and X satisfy 2.03 for *all* A. Then, taking 0 and X in turn for A, and using 2.01, we have

$$0 = 0 \cup X = X \cup 0 = X$$

proving that $X = 0$.
Similarly if 1 and Y satisfy 2.03, taking 1 and Y in turn for A, and using 2.01, we have

$$1 = 1 \cap Y = Y \cap 1 = Y$$

so that $Y = 1$. Thus 0 and 1 are unique classes. Furthermore if A has two complements A', A^* both satisfying 2.04 then

$$
\begin{aligned}
A^* &= A^* \cup 0 && \text{by 2.03} \\
&= A^* \cup (A \cap A') && \text{by 2.04} \\
&= (A^* \cup A) \cap (A^* \cup A') && \text{by 2.02} \\
&= (A \cup A^*) \cap (A^* \cup A') && \text{by 2.01} \\
&= 1 \cap (A^* \cup A') && \text{by 2.04} \\
&= (A^* \cup A') \cap 1 && \text{again by 2.01} \\
&= A^* \cup A' && \text{by 2.03.}
\end{aligned}
$$

Similarly $A' = A' \cup A^* = A^* \cup A' = A^*$ which proves that $A^* = A'$. Thus the classes 0, 1 are unique and so is the complement of a class.

(Note that although there is only one class 0 such that $A \cup 0 = A$, for *all* A, there are other classes X such that $A \cup X = A$ for an X depending on A, for instance A itself is one such class as we show below.)

2.21. Next we show that complementation is involutory, that is, that the complement of the complement is the original class. For from 2.04 and 2.01

$$A' \cup A = 1, \quad A' \cap A = 0$$

which show that A is the complement of A'; but the complement is unique, and so $A'' = A$.

2.22. In the same way we show that 0, 1 are each the complement of the other. For, from 2.03, and 2.01,

$$1 \cup 0 = 1, \qquad 1 \cap 0 = 0$$

and from 2.04, and the uniqueness of the complement, it now follows that $0 = 1'$; consequently $0' = 1'' = 1$.

2.3. Two of the properties we established in section 1.5 we have taken as basic in 2.03, and there remain to be proved

2.31. $\qquad\qquad A \cup 1 = 1, \qquad A \cap 0 = 0$

2.32. $\qquad\qquad A \cup A = A, \qquad A \cap A = A.$

For the first of these we have

$$\begin{aligned}
A \cup 1 &= (A \cup 1) \cap 1 = 1 \cap (A \cup 1) \\
&= (A \cup A') \cap (A \cup 1) \\
&= A \cup (A' \cap 1) \\
&= A \cup A' \\
&= 1.
\end{aligned}$$

Similarly

$$\begin{aligned}
A \cap 0 &= (A \cap 0) \cup 0 = 0 \cup (A \cap 0) \\
&= (A \cap A') \cup (A \cap 0) \\
&= A \cap (A' \cup 0) \\
&= A \cap A' \\
&= 0.
\end{aligned}$$

To prove the so-called idempotent laws 2.32 we proceed as follows:

$$\begin{aligned}
A = A \cup 0 = A \cup (A \cap A') &= (A \cup A) \cap (A \cup A') \\
&= (A \cup A) \cap 1 \\
&= A \cup A;
\end{aligned}$$

$$\begin{aligned}
A = A \cap 1 = A \cap (A \cup A') &= (A \cap A) \cup (A \cap A') \\
&= (A \cap A) \cup 0 \\
&= A \cap A.
\end{aligned}$$

2.4. We prove next the two *absorption* laws

$$A \cup (A \cap B) = A, \qquad A \cap (A \cup B) = A.$$

For the first of these we have

$$
\begin{aligned}
A \cup (A \cap B) &= (A \cap 1) \cup (A \cap B) \\
&= A \cap (1 \cup B), && \text{by the distributive law,} \\
&= A \cap 1, && \text{using 2.01 and 2.31,} \\
&= A,
\end{aligned}
$$

and similarly

$$A \cap (A \cup B) = (A \cup 0) \cap (A \cup B)$$

$$= A \cup (0 \cap B) = A \cup 0 = A.$$

2.5. **Duality.** An examination of the properties of classes we took as basic, in 2.01–2.04, shows that these properties express a duality between the operations of union and intersection; we set out these properties in pairs to exhibit this fact. The two relations 2.01 each transform into the other by interchanging $\cup$ and $\cap$, and the same is true of the two relations in 2.02. In 2.03 and 2.04 this interchange is valid provided that we also exchange 0 and 1 at the same time. It follows that in any relation derived from these basic properties the interchange of $\cup$ and $\cap$ and of 0, 1 simultaneously gives a relation which is also derivable from the same properties—to prove the second we have only to carry out these changes in the proof of the first. This is readily seen in the proofs we have given of the pairs of relations 2.31, 2.32 and 2.4. In each case the proof of the second relation is obtained from that of the first just by making these changes. In the sequel we shall no longer write out both proofs but shall content ourselves with calling attention to the dual relation.

2.51. In the next two theorems we give two important ways of proving the equality of classes. The first of these is:

2.52.   if, for some A, B, C,

$$A \cup B = A \cup C, \quad \text{and} \quad A \cap B = A \cap C$$

then $\qquad\qquad\qquad B = C.$

For

$$\begin{aligned}
B = B \cap (B \cup A) &= B \cap (C \cup A) \\
&= (B \cap C) \cup (B \cap A) \\
&= (B \cap C) \cup (C \cap A) \\
&= C \cap (A \cup B) \\
&= C \cap (A \cup C) = C
\end{aligned}$$

proving that $\qquad\qquad B = C.$

The second result is that:

2.53.   if, for some A, B, C,

$$A \cup B = A \cup C, \quad A' \cup B = A' \cup C$$

then $\qquad\qquad\qquad B = C.$

For $\quad (A \cup B) \cap (A' \cup B) = B \cup (A \cap A') = B \cup 0 = B$

and

$$(A \cup C) \cap (A' \cup C) = C \cup (A \cap A') = C \cup 0 = C,$$

whence the result follows.

The dual result is that from

$$A \cap B = A \cap C, \quad A' \cap B = A' \cap C$$

follows

$$B = C.$$

2.6. It is important to notice that we did not include the associative laws amongst the basic assumptions .01–.04. The result we have just established, however, may be used to prove the associative laws for both union and intersection. It will, of course, be necessary to prove only one of these laws since the other follows by duality. Let us consider the associative law for union, namely

$$A \cup (B \cup C) = (A \cup B) \cup C,$$

and let us write $L = A \cup (B \cup C)$ and $M = (A \cup B) \cup C$.

Then $A \cap L = A \cap (A \cup (B \cup C)) = A,$ by 2.4

and $A \cap M = (A \cap (A \cup B)) \cup (A \cap C)$
$$= A \cup (A \cap C) = A, \qquad \text{by 2.4 again,}$$
so that $A \cap L = A \cap M.$
Furthermore

$$
\begin{aligned}
A' \cap L &= A' \cap (A \cup (B \cup C)) \\
&= (A' \cap A) \cup (A' \cap (B \cup C)) \\
&= 0 \cup (A' \cap (B \cup C)) \\
&= A' \cap (B \cup C) = (A' \cap B) \cup (A' \cap C),
\end{aligned}
$$

and $A' \cap M = A' \cap ((A \cup B) \cup C)$
$$
\begin{aligned}
&= (A' \cap (A \cup B)) \cup (A' \cap C) \\
&= ((A' \cap A) \cup (A' \cap B)) \cup (A' \cap C) \\
&= (0 \cup (A' \cap B)) \cup (A' \cap C) \\
&= (A' \cap B) \cup (A' \cap C)
\end{aligned}
$$

so that
$$A' \cap L = A' \cap M$$

and therefore, by .53, we have $L = M$, which proves the associative property for union.

2.601. A simple consequence of associativity is that if for some classes A, B

$$A \cup B = A \cap B$$

then $A = B.$

For

$$A = A \cap (A \cup B) = A \cap (A \cap B) = (A \cap A) \cap B = A \cap B$$

and

$$B = B \cap (A \cup B) = B \cap (A \cap B) = (B \cap B) \cap A = B \cap A$$

whence, since $A \cap B = B \cap A$, we have $A = B.$

2.61. We come now to the De Morgan laws

$$(A \cup B)' = A' \cap B', \qquad (A \cap B)' = A' \cup B'.$$

Consider

$$
\begin{aligned}
(A \cup B) \cup (A' \cap B') &= (A \cup B \cup A') \cap (A \cup B \cup B'), \quad \text{by .02,} \\
&= ((A \cup A') \cup B) \cap (A \cup (B \cup B')) \\
&= (1 \cup B) \cap (A \cup 1) \\
&= 1 \cap 1 = 1
\end{aligned}
$$

and

$$
\begin{aligned}
(A \cup B) \cap (A' \cap B') &= (A' \cap B' \cap A) \cup (A' \cap B' \cap B) \\
&= ((A \cap A') \cap B') \cup (A' \cap (B \cap B')) \\
&= (0 \cap B') \cup (A' \cap 0) \\
&= 0 \cup 0 = 0;
\end{aligned}
$$

since the complement is unique, and $A' \cap B'$ satisfies the two defining characteristics .04 of the complement, it follows that $A' \cap B'$ *is* the complement of $A \cup B$, as was to be proved. The second result follows by duality.

2.62. **Inclusion.** We proved in section 1.62 that $A \subset B$ if and only if $A = A \cap B$ and we now take this equivalence as *defining* the relation of inclusion. That is, *we define $A \subset B$* (read *A is contained in B*) *to stand for* $A = A \cap B$.

We start by remarking that the equation $A = A \cap B$ is equivalent to $B = A \cup B$ for if $A = A \cap B$ then $B = B \cup (B \cap A) = B \cup A = A \cup B$, and conversely if $B = A \cup B$ then

$$A = A \cap (A \cup B) = A \cap B.$$

Immediate consequences of the definition are, for all A,

$$0 \subset A, \qquad A \subset 1$$

for $0 \cap A = 0$, $1 \cap A = A$.

2.63. Inclusion is transitive, that is to say,

if
$$A \subset B \quad \text{and} \quad B \subset C$$

then $\qquad\qquad\qquad A \subset C.$

We have to show that if

$$A = A \cap B \quad \text{and} \quad B = B \cap C \quad \text{then} \quad A = A \cap C.$$

We have $A \cap C = (A \cap B) \cap C = A \cap (B \cap C) = A \cap B = A$.
2.64. The condition $A \subset B$ is equivalent to $A \cap B' = 0$.
For if $A \subset B$ then $A = A \cap B$ and so $A \cap B' = A \cap (B \cap B')$
$= A \cap 0 = 0$.
Conversely, if $A \cap B' = 0$ then

$$\begin{aligned} A = A \cap 1 = A \cap (B \cup B') &= (A \cap B) \cup (A \cap B') \\ &= (A \cap B) \cup 0 = A \cap B. \end{aligned}$$

It follows that $A \subset B$ is equivalent to $A' \cup B = 1$.
2.65. If $A \subset B$ and $B \subset A$ then $A = B$; for from $A \subset B$ and
$B \subset A$ follows

$$A = A \cap B = B \cap A = B.$$

Conversely, if $A = B$ then $A \subset B$, for from $A = B$ follows

$$A = A \cap A = A \cap B.$$

In particular $A \subset A$, because $A = A \cap A$.
2.66. For any classes A, B

$$A \cap B \subset A, \qquad A \subset A \cup B.$$

We have $(A \cap B) \cap A = (A \cap A) \cap B = A \cap B$ which proves
that $A \cap B \subset A$; and $A \cup (A \cup B) = (A \cup A) \cup B = A \cup B$
which proves that $A \subset A \cup B$.
2.67. We list now a succession of relations which we proved in
the previous chapter by means of properties which we have now
established in this chapter. Proofs are identical with those given
before and will not be repeated; instead we give the section number
for reference.

If $A \subset B$ and $A \subset C$ then $A \subset B \cap C$ $\qquad$ (1.82)

If $A \subset C$ and $B \subset C$ then $A \cup B \subset C$ $\qquad$ (1.82)

If $A \subset B$ then $A \cap C \subset B \cap C$ and $A \cup C \subset B \cup C$ $\qquad$ (1.83)

If $B \subset A$ for all A, then $B = 0$ $\qquad$ (1.86)

If $A \subset B$ for all A, then $B = 1$ $\qquad$ (1.86)

Defining the difference $A - B$ to equal $A \cap B'$ we have

$$1 - A = A' \tag{1.9}$$

$$A - B = 0 \quad \text{is equivalent to} \quad A \subset B \tag{1.91}$$

$$(A - B) \cup B = A \cup B \tag{1.92}$$

and so, if $B \subset A$ then $(A - B) \cup B = A$.

If $A \cap B = 0$ then $A - B = A$ $\qquad$ (1.92)

$$C \cap (A - B) = (C \cap A) - (C \cap B) \tag{1.93}$$

Defining the symmetric difference $A + B$ as $(A - B) \cup (B - A)$ we have

$$A + B = (A \cup B) \cap (A' \cup B') \tag{1.94}$$

$$A + B = B + A \tag{1.95}$$

$$\begin{aligned}
(A + B) + C = A + (B + C) &= (A \cup B \cup C) \cap (A' \cup B' \cup C) \cap \\
&\quad (A \cup B' \cup C') \cap (A' \cup B \cup C') \\
&= (A \cap B \cap C) \cup (A' \cap B' \cap C) \cup \\
&\quad (A' \cap B \cap C') \cup (A \cap B' \cap C').
\end{aligned}$$

$\qquad$ (1.95)

We prove this last result in detail since the proof in 1.99 is by a Venn diagram.

We have $(A + B) + C$

$$\begin{aligned}
&= \{((A \cap B') \cup (A' \cap B)) \cap C'\} \cup \{((A \cap B') \cup (A' \cap B))' \cap C\} \\
&= \{((A \cap B') \cup (A' \cap B)) \cap C'\} \cup \{(A' \cup B) \cap (A \cup B') \cap C\} \\
&= (A \cap B' \cap C') \cup (A' \cap B \cap C') \cup (A' \cap A \cap C) \\
&\quad \cup (A' \cap B' \cap C) \cup (B \cap A \cap C) \cup (B \cap B' \cap C) \\
&= (A \cap B \cap C) \cup (A' \cap B' \cap C) \cup (A' \cap B \cap C') \cup (A \cap B' \cap C').
\end{aligned}$$

2.7. Standard forms. Any expression built out of classes A, B, C, ... , by means of union, intersection and complementation may be reduced to the standard form

$$U_1 \cap U_2 \cap U_3 \cap ... \cap U_k$$

where each U is a union of classes or complement classes. To effect this reduction we use the De Morgan laws to replace the complement of a union or intersection by the intersection or union of complements, and we use the distributive law to distribute union over intersection. In other words we take complementation *into* a bracket, and open intersections separated by unions. To illustrate the reduction we consider the expression

$$(A \cap B \cap C) \cup ((C \cup D)' \cup (D' \cap E))';$$

by De Morgan's laws we transform this in turn into

$$(A \cap B \cap C) \cup ((C \cup D) \cap (D \cup E'))$$
$$= (A \cup C \cup D) \cap (A \cup D \cup E') \cap (B \cup C \cup D) \cap (B \cup D \cup E') \cap$$
$$(C \cup D) \cap (C \cup D \cup E')$$

which is in standard form.

A second standard form to which any expression may be reduced is

$$I_1 \cup I_2 \cup ... \cup I_k$$

where each I is an intersection of elements or complemented elements. This form may be obtained exactly as above, but using the distributive law to distribute intersection over union; or we may first transform the complement of the given expression to the form

$$U_1 \cap U_2 \cap U_3 \cap ... \cap U_k$$

the complement of which is the second standard form (because the complement of an intersection is a union, and the complement of a union is an intersection).

For instance the complement of

$$(A \cup C \cup D) \cap (A \cup D \cup E') \cap (B \cup C \cup D) \cap (B \cup D \cup E')$$
$$\cap (C \cup D) \cap (C \cup D \cup E')$$

is

$$(A' \cap C' \cap D') \cup (A' \cap D' \cap E) \cup (B' \cap C' \cap D') \cup (B' \cap D' \cap E)$$
$$\cup (C' \cap D') \cup (C' \cap D' \cap E).$$

Either of the two standard forms provides a purely mechanical procedure for determining whether an expression is equal to zero, or not. For instance an expression with the standard form

$$I_1 \cup I_2 \cup \ldots \cup I_k$$

is equal to zero if and only if *each* $I_1, I_2, \ldots, I_k$ is equal to zero, and an intersection of elements or complemented elements is equal to zero *if and only if some element and its complement are both present.*
Thus

$$(A \cap B \cap A') \cup (B \cap C \cap C')$$

is equal to zero, being the union of intersections of some class with its complement.

For instance to prove that

$$(A' \cap B)' \cap (A \cap C)' \cap B \cap C = 0$$

we transform the left-hand side successively into

$$(A \cup B') \cap (A' \cup C') \cap B \cap C$$

$$= ((A \cap B) \cup (B' \cap B)) \cap ((A' \cap C) \cup (C' \cap C))$$

$$= ((A \cap B) \cap (A' \cap C)) \cup ((A \cap B) \cap C \cap C')$$

$$\cup ((B \cap B') \cap A' \cap C) \cup (B \cap B' \cap C \cap C')$$

$$= (A \cap A' \cap B \cap C) \cup (A \cap B \cap C \cap C') \cup (B \cap B' \cap A' \cap C)$$

$$\cup (B \cap B' \cap C \cap C')$$

$$= 0.$$

(Of course, in this example we might have shortened the proof by replacing $B \cap B'$ and $C \cap C'$ by 0 earlier in the proof.)

2.71. If we denote any *expression*, built up from classes $A_1, A_2,$ ... , A_n by complementation, union and intersection, by $f(A_1, A_2,$ ... , $A_n)$ then by $f(A_1, A_2, ... , A_{r-1}, 0, A_{r+1}, ... , A_n)$ we denote the expression obtained from $f(A_1, ... , A_n)$ by writing 0 for A_r wherever it occurs, and similarly we denote by $f(A_1, A_2, ... , A_{r-1}, 1, A_{r+1}, ... , A_n)$ the expression obtained by writing 1 for A_r.

We shall show that every expression $f(A_1, A_2, ... , A_n)$ may be obtained from $A_1, ... , A_n$ and their complements, associated with expressions involving 0, 1 alone.

Consider first an expression $f(A)$ containing a single class A. Then we prove that

2.72. $$f(A) = \{A \cup f(0)\} \cap \{A' \cup f(1)\}.$$

Observe first that if $f(A) = A$ then this relation becomes

$$A = (A \cup 0) \cap (A' \cup 1)$$

which is certainly true.

Next observe that if .72 holds for some expression $f(A)$ it holds for the complement of this expression, for the complement of $\{A \cup f(0)\} \cap \{A' \cup f(1)\}$ is

$$\{A' \cap f'(0)\} \cup \{A \cap f'(1)\}$$

$= \{A \cup f'(0)\} \cap \{A' \cup f'(1)\}$, by example II,   1.92.

Moreover if .72 holds for two expressions $f(A), g(A)$ it holds for their union since
$$[\{A \cup f(0)\} \cap \{A' \cup f(1)\}] \cup [\{A \cup g(0)\} \cap \{A' \cup g(1)\}]$$
$= \{A \cup f(0) \cup g(0)\} \cap \{A' \cup f(1) \cup g(1)\}$ (see example

II,1.4).

Finally, .72 holds for the intersection of $f(A), g(A)$ which satisfy .72, since
$$\{A \cup f(0)\} \cap \{A' \cup f(1)\} \cap \{A \cup g(0)\} \cap \{A' \cup g(1)\}$$
$$= \{A \cup (f(0) \cap g(0))\} \cap \{A' \cup (f(1) \cap g(1))\}.$$

Since every expression is built up by complementation, union and intersection, it follows that .72 holds for *any* expression $f(A)$.

Notice that .72 holds even when $f(A)$ is some expression K which does not contain A at all, so that $f(0) = f(1) = K$, for

$$(A \cup K) \cap (A' \cup K) = (A \cap A') \cup K = K.$$

Consider now an expression $f(A_1, A_2, \ldots, A_n, A)$ built up from classes $A_1, A_2, \ldots, A_n$ and A; we have, exactly as before,

2.73. $f(A_1, A_2, \ldots, A_n, A)$
$$= \{A \cup f(A_1, A_2, \ldots, A_n, 0)\} \cap \{A' \cup f(A_1, A_2, \ldots, A_n, 1)\}$$

for this relation holds if $f(A_1, \ldots, A_n, A)$ does not in fact contain A, so that

$$f(A_1, \ldots, A_n, A) = f(A_1, A_2, \ldots, A_n, 0) = f(A_1, A_2, \ldots, A_n, 1) = K,$$

say; it holds when $f(A_1, \ldots, A_n, A) = A \cup g(A_1, A_2, \ldots, A_n)$ or $A \cap g(A_1, A_2, \ldots, A_n)$ (since $g(A_1, A_2, \ldots, A_n)$ does not contain A), and it holds for the complement, union and intersection of expressions for which it is known to hold. Thus .73 holds for all expressions.

From .73, we may step by step express any expression in the desired form. First we observe that .72 expresses $f(A)$ in terms of A, 0 and 1. Then using .72 and .73 we have

$$f(A, B) = \{A \cup f(0, B)\} \cap \{A' \cup f(1, B)\}; \quad \text{but}$$
$$f(0, B) = \{B \cup f(0, 0)\} \cap \{B' \cup f(0, 1)\},$$
$$f(1, B) = \{B \cup f(1, 0)\} \cap \{B' \cup f(1, 1)\} \quad \text{and so}$$
$$f(A, B) = \{A \cup B \cup f(0, 0)\} \cap \{A \cup B' \cup f(0, 1)\}$$
$$\cap \{A' \cup B \cup f(1, 0)\} \cap \{A' \cup B' \cup f(1, 1)\}$$

and similarly

$$f(A, B, C) = \{A \cup B \cup C \cup f(0, 0, 0)\} \cap \{A \cup B \cup C' \cup f(0, 0, 1)\}$$
$$\cap \{A \cup B' \cup C \cup f(0, 1, 0)\} \cap \{A' \cup B \cup C \cup f(1, 0, 0)\}$$
$$\cap \{A \cup B' \cup C' \cup f(0, 1, 1)\} \cap \{A' \cup B \cup C' \cup f(1, 0, 1)\}$$
$$\cap \{A' \cup B' \cup C \cup f(1, 1, 0)\} \cap \{A' \cup B' \cup C' \cup f(1, 1, 1)\},$$

and so on.

Of course each expression $f(U_1, U_2, \ldots, U_n)$, where each U is 0 or 1, is equal to 0 or to 1. Thus we obtain all distinct (i.e. unequal) expressions in A by taking $f(0), f(1)$ to be 0 or 1 in all possible ways in .72 giving, in all, the following expressions:

$$(A \cup 0) \cap (A' \cup 0) = A \cap A' = 0$$
$$(A \cup 0) \cap (A' \cup 1) = A \cap 1 = A$$
$$(A \cup 1) \cap (A' \cup 0) = 1 \cap A' = A'$$
$$(A \cup 1) \cap (A' \cup 1) = 1 \cap 1 = 1$$

which shows that every expression in A alone is equal to one of 0, 1, A and A'.

Similarly we see that every expression in A, B alone is one of the 16 expressions

$$\{A \cup B \cup 0\} \cap \{A \cup B' \cup 0\} \cap \{A' \cup B \cup 0\} \cap \{A' \cup B' \cup 0\} = 0$$
$$\{A \cup B \cup 0\} \cap \{A \cup B' \cup 0\} \cap \{A' \cup B \cup 0\} \cap \{A' \cup B' \cup 1\} = A \cap B$$
$$\{A \cup B \cup 0\} \cap \{A \cup B' \cup 0\} \cap \{A' \cup B \cup 1\} \cap \{A' \cup B' \cup 0\} = A - B$$
$$\{A \cup B \cup 0\} \cap \{A \cup B' \cup 0\} \cap \{A' \cup B \cup 1\} \cap \{A' \cup B' \cup 1\} = A$$
$$\{A \cup B \cup 0\} \cap \{A \cup B' \cup 1\} \cap \{A' \cup B \cup 0\} \cap \{A' \cup B' \cup 0\} = B - A$$
$$\{A \cup B \cup 0\} \cap \{A \cup B' \cup 1\} \cap \{A' \cup B \cup 0\} \cap \{A' \cup B' \cup 1\} = B$$
$$\{A \cup B \cup 0\} \cap \{A \cup B' \cup 1\} \cap \{A' \cup B \cup 1\} \cap \{A' \cup B' \cup 0\} = A + B$$
$$\{A \cup B \cup 0\} \cap \{A \cup B' \cup 1\} \cap \{A' \cup B \cup 1\} \cap \{A' \cup B' \cup 1\} = A \cup B$$
$$\{A \cup B \cup 1\} \cap \{A \cup B' \cup 0\} \cap \{A' \cup B \cup 0\} \cap \{A' \cup B' \cup 0\} = A' \cap B'$$
$$\{A \cup B \cup 1\} \cap \{A \cup B' \cup 0\} \cap \{A' \cup B \cup 0\} \cap \{A' \cup B' \cup 1\} = (A + B)'$$
$$\{A \cup B \cup 1\} \cap \{A \cup B' \cup 0\} \cap \{A' \cup B \cup 1\} \cap \{A' \cup B' \cup 0\} = B'$$
$$\{A \cup B \cup 1\} \cap \{A \cup B' \cup 0\} \cap \{A' \cup B \cup 1\} \cap \{A' \cup B' \cup 1\} = A \cup B'$$
$$\{A \cup B \cup 1\} \cap \{A \cup B' \cup 1\} \cap \{A' \cup B \cup 0\} \cap \{A' \cup B' \cup 0\} = A'$$
$$\{A \cup B \cup 1\} \cap \{A \cup B' \cup 1\} \cap \{A' \cup B \cup 0\} \cap \{A' \cup B' \cup 1\} = A' \cup B$$
$$\{A \cup B \cup 1\} \cap \{A \cup B' \cup 1\} \cap \{A' \cup B \cup 1\} \cap \{A' \cup B' \cup 0\} = A' \cup B'$$
$$\{A \cup B \cup 1\} \cap \{A \cup B' \cup 1\} \cap \{A' \cup B \cup 1\} \cap \{A' \cup B' \cup 1\} = 1.$$

2.74. It follows immediately from the last section that *if two expressions* $f(A_1, A_2, \ldots, A_n)$, $g(A_1, A_2, \ldots, A_n)$ *are equal for the values 0, 1 of each* A_i *then the equation* $f(A_1, A_2, \ldots, A_n) = g(A_1, A_2, A_n)$ *is provable from 2.01–2.04.* Consider for example the case of two classes. By hypothesis $f(0, 0) = g(0, 0)$, $f(0, 1) = g(0, 1)$, $f(1, 0) = g(1, 0)$ and $f(1, 1) = g(1, 1)$, and therefore

$$\begin{aligned}
f(A, B) &= \{A \cup B \cup f(0, 0)\} \cap \{A \cup B' \cup f(0, 1)\}\\
&\quad \cap \{A' \cup B \cup f(1, 0)\} \cap \{A' \cup B' \cup f(1, 1)\}\\
&= \{A \cup B \cup g(0, 0)\} \cap \{A \cup B' \cup g(0, 1)\}\\
&\quad \cap \{A' \cup B \cup g(1, 0)\} \cap \{A' \cup B' \cup g(1, 1)\}\\
&= g(A, B).
\end{aligned}$$

Consequently *every relation of the algebra of classes which can be established by the methods of Chapter I, is provable from 2.01–2.04.* For a relation

$$f(A_1, A_2, \ldots, A_n) = g(A_1, A_2, \ldots, A_n)$$

which is true for any classes $A_1, A_2, \ldots, A_n$ is true in particular for the classes 0, 1 and so is provable from 2.01–2.04. We express this result by saying that the properties 2.01–2.04 and their consequences are *complete* with respect to their interpretation as an algebra of classes. We have also incidentally found *a purely mechanical test for deciding whether a relation is provable or not;* we have only to test it for the two classes 0, 1. For example to see whether the relation

$$(A \cup B') \cap B = A \cap B$$

is provable or not we set up the table

A	B	B'	$A \cup B'$	$(A \cup B') \cap B$	$A \cap B$
0	0	1	1	0	0
0	1	0	0	0	0
1	0	1	1	0	0
1	1	0	1	1	1

In the first two columns we enter the four possible combinations of values 0, 1 for A and B. In column three, in each row we enter the complement of the value of B in the same row; in column four we enter the union of the values of A and B' in each row (thus in the first row we enter $0 \cup 1 = 1$), and so on. The fact that columns five and six are the same shows that the equation is true, and so provable.

2.8. The basic relations 2.01–2.04 form what is called an *axiom system for Boolean algebra*; this particular system of axioms was given by E. V. Huntington in 1904. The name "Boolean algebra" derives from G. Boole who first introduced the fundamental concepts and properties of this algebra in his books *The Mathematical Analysis of Logic* (1847) and *The Laws of Thought* (1854). We have seen, in the previous chapter, that the axioms 2.01–2.04 express true sentences if we interpret the variables $A, B, C, \ldots$ as standing for classes, the elements 0, 1 standing for the null and the universal class, and A' standing for the complement of A in 1. For this reason we say that the algebra of classes is a *model* for the axioms 2.01–2.04. Another model for the axioms, a model of a very different kind, is given by the factors of 30. If we give the variables $A, B, C, \ldots$ the values 1, 2, 3, 5, 6, 10, 15, 30 and if we interpret $A \cup B$ to mean the least common multiple of A, B and $A \cap B$ to mean the highest common factor (with **0** standing for 1 and **1** standing for 30) the relations 2.01–2.04 again express truths, truths of arithmetic, as may readily be verified.†

Formally we define a Boolean algebra as a set Σ of at least two elements, including 0, 1, which admits three operations, called complementation, union and intersection and denoted by A', $A \cup B$, $A \cap B$ respectively, such that for all elements A, B of Σ A', $A \cup B$, $A \cap B$ are elements of Σ which satisfy 2.01–2.04. For any A, B of Σ, A', $A \cup B$, $A \cap B$ belong to Σ and so Σ is said to be *closed* under complementation, union and intersection. In the axioms the letters A, B, C stand for any elements of the set Σ, and not just for a particular three elements. There are various

† See example II, 10.

ways in which the generality of the axioms can be expressed. We may suppose that 2.01–2.04 are not the actual axioms but are *matrices* out of which axioms are formed by taking actual elements in place of Σ; alternatively we may treat the elements of Σ, if there are others than just 0 and 1, as *variables* for which elements may be substituted. In the first case A, B, C are not themselves elements and the axioms are unlimited in number, each instance of 2.01–2.04 with elements in place of A, B, C, being an axiom. In the second case A, B, C are themselves elements and 2.01–2.04 are the only axioms, but now we require a rule of substitution which allows us to *substitute* elements for A, B, C. We shall have no occasion to maintain a distinction between these two cases, but when one deals with specific Boolean algebras, e.g. the Boolean algebra with elements which are the subclasses of the class (a, b, c), including the null class 0 and the complete class $1 = (a, b, c)$, then the axioms must be expressed by means of variables which are not themselves elements of the algebra.

The basic properties 2.01–2.04, or axioms as we shall now call them, are independent of one another, that is to say not one of them is dispensable, and not one of them is a consequence of the others. To show this we shall exhibit models which satisfy all but one of the axioms and all the consequences of all but this one axiom.

2.81. We consider first the algebra with just two elements 0, 1, each the complement of the other, and operations $\cup$, $\cap$ defined by the tables

$\cup$	0	1
0	0	1
1	1	0

$\cap$	0	1
0	0	0
1	0	1

In this algebra all the axioms are valid except the first distributive law. That axioms 2.01 are satisfied is evident since the tables are symmetrical about their leading diagonals. To prove 2.03 we have only to verify $1 \cup 0 = 1$, $0 \cup 0 = 0$, $1 \cap 1 = 1$, $0 \cap 1 = 0$ as is immediately seen to be true from the tables. Axioms 2.04 similarly require $0 \cup 1 = 1 \cup 0 = 1$ and $0 \cap 1 = 1 \cap 0 = 0$. For the second distributive law we consider separately the two cases $A = 0$, $A = 1$.

For $A = 0$, we have

$$0 \cap (B \cup C) = 0 = 0 \cap B = 0 \cap C = (0 \cap B) \cup (0 \cap C)$$

and for $A = 1$,

$$1 \cap (B \cup C) = B \cup C, \qquad 1 \cap B = B, \qquad 1 \cap C = C,$$

$$(1 \cap B) \cup (1 \cap C) = B \cup C,$$

which proves that the second distributive law holds. But the first distributive law fails because

$$1 \cup (1 \cap 0) = 1$$

but $\qquad\qquad (1 \cup 1) \cap (1 \cup 0) = 1 \cap 0 = 0.$

2.82. An algebra which exhibits the independence of the second distributive law is again an algebra with just two elements 0, 1, each of which is the complement of the other, in which $\cap$ and $\cup$ are defined by the tables

$\cup$	0	1
0	0	1
1	1	1

$\cap$	0	1
0	1	0
1	0	1

We shall omit the detailed verification of all the axioms except the second distributive law; this law is not satisfied because

$$0 \cap (0 \cup 1) = 0 \cap 1 = 0$$

but

$$(0 \cap 0) \cup (0 \cap 1) = 1 \cup 0 = 1.$$

For the commutative laws 2.01 we may prove independence by the models

2.83.

$\cup$	0	1
0	0	0
1	1	1

$\cap$	0	1
0	0	0
1	0	1

and

2.84.

$\cup$	0	1
0	0	1
1	1	1

$\cap$	0	1
0	0	0
1	1	1

The first model exhibits the independence of the axiom

$$A \cup B = B \cup A$$

and the second the independence of

$$A \cap B = B \cap A.$$

Again we leave to the reader the verification of the valid axioms and content ourselves with exhibiting the failure of an axiom for its appropriate model. In the present instance, the failure is obvious since in the first model

$$0 \cup 1 = 0, \quad 1 \cup 0 = 1$$

and in the second

$$0 \cap 1 = 0, \quad 1 \cap 0 = 1.$$

Since, however, in 2.83 both 0 and 1 satisfy $A \cup 0 = A$ (i.e. the zero is not unique) we must ignore axioms 2.04, and in 2.84 the unit has no complement and is not unique.

A model which exhibits the independence of the axiom

$$A \cup 0 = A$$

whilst preserving the remaining axioms is

2.85.

$\cup$	a	b
a	a	a
b	a	a

$\cap$	a	b
a	a	a
b	a	b

In this model *no* element plays the part of zero, and so axioms 2.04 cannot be considered.

A model for the independence of the axiom

$$A \cap 1 = A$$

is

2.86.

$\cup$	a	b
a	a	b
b	b	b

$\cap$	a	b
a	b	b
b	b	b

In this model there is no unit, and so again axiom 2.04 is ignored.

To prove the independence of the axioms for complements we may use the model

$\cup$	a	b
a	a	b
b	b	b

$\cap$	a	b
a	a	b
b	b	b

Here a plays the part of both 0 and 1, and b has no complement.

Finally we observe that *all* the axioms are satisfied if there is a *single* class a satisfying

$$a \cup a = a, \qquad a \cap a = a,$$

and this shows why we lay down the requirement that there are at least two classes, that is, we require the zero and unit to be distinct.

An algebra of pairs

2.9. In this section we set up a Boolean algebra of pairs of elements of a Boolean algebra.

We start by defining the complement, union and intersection of pairs.

The complement of (A, B) is defined to be (A', B').

The union of (A_1, B_1), (A_2, B_2) is

$$(A_1, B_1) \cup (A_2, B_2) = (A_1 \cup A_2, B_1 \cap B_2)$$

and their intersection

$$(A_1, B_1) \cap (A_2, B_2) = (A_1 \cap A_2, B_1 \cup B_2).$$

We remark first that axioms 2.01 are clearly satisfied, and proceed to verify the distributive laws. We have

$$
\begin{aligned}
(A_1, B_1) &\cup \{(A_2, B_2) \cap (A_3, B_3)\} \\
&= (A_1, B_1) \cup (A_2 \cap A_3, B_2 \cup B_3) \\
&= (A_1 \cup (A_2 \cap A_3), B_1 \cap (B_2 \cup B_3)) \\
&= ((A_1 \cup A_2) \cap (A_1 \cup A_3), (B_1 \cap B_2) \cup (B_1 \cap B_3))
\end{aligned}
$$

and

$$
\begin{aligned}
\{(A_1, B_1) &\cup (A_2, B_2)\} \cap \{(A_1, B_1) \cup (A_3, B_3)\} \\
&= (A_1 \cup A_2, B_1 \cap B_2) \cap (A_1 \cup A_3, B_1 \cap B_3) \\
&= ((A_1 \cup A_2) \cap (A_1 \cup A_3), (B_1 \cap B_2) \cup (B_1 \cap B_3))
\end{aligned}
$$

which proves the first of 2.02. For the second we find

$$
\begin{aligned}
(A_1, B_1) &\cap \{(A_2, B_2) \cup (A_3, B_3)\} \\
&= (A_1, B_1) \cap (A_2 \cup A_3, B_2 \cap B_3) \\
&= (A_1 \cap (A_2 \cup A_3), B_1 \cup (B_2 \cap B_3)) \\
&= ((A_1 \cap A_2) \cup (A_1 \cap A_3), (B_1 \cup B_2) \cap (B_1 \cup B))_3
\end{aligned}
$$

and

$$((A_1, B_1) \cap (A_2, B_2)) \cup ((A_1, B_1) \cap (A_3, B_3))$$
$$= (A_1 \cap A_2, B_1 \cup B_2) \cup (A_1 \cap A_3, B_1 \cup B_3)$$
$$= ((A_1 \cap A_2) \cup (A_1 \cap A_3), (B_1 \cup B_2) \cap (B_1 \cup B_3))$$

proving the second distributive law.

The zero and unit pairs are respectively

$$(0, 1) \quad \text{and} \quad (1, 0).$$

We have

$$(A, B) \cup (0, 1) = (A \cup 0, B \cap 1) = (A, B)$$

and

$$(A, B) \cap (1, 0) = (A \cap 1, B \cup 0) = (A, B)$$

verifying 2.03, and

$$(A, B) \cup (A, B)' = (A, B) \cup (A', B') = (A \cup A', B \cap B') = (1, 0),$$
$$(A, B) \cap (A, B)' = (A, B) \cap (A', B') = (A \cap A', B \cup B') = (0, 1),$$

completing the verification of all the axioms.

2.91. If we select the subset of pairs (A, A') from the set of pairs (A, B) we observe that, in virtue of what we have just proved, this set itself constitutes a Boolean algebra (since the zero and unit pairs $(0, 1)$, $(1, 0)$ are themselves of this form). There is a one-to-one correspondence between the elements A and the pairs (A, A'), for to each element A corresponds exactly one pair (A, A') and to each pair (A, A') corresponds its first term A. We express this correspondence by writing

$$A \leftrightarrow (A, A').$$

This one-to-one correspondence is preserved under complementation, union and intersection, and the zeros and units are married in the correspondence. Thus

$$A \leftrightarrow (A, A'), \qquad A' \leftrightarrow (A', A) = (A, A')'$$
$$A \cup B \leftrightarrow (A \cup B, A' \cap B') = (A, A') \cup (B, B')$$
$$A \cap B \leftrightarrow (A \cap B, A' \cup B') = (A, A') \cap (B, B')$$

and

$$0 \leftrightarrow (0, 1), \qquad 1 \leftrightarrow (1, 0).$$

A one-to-one correspondence between two sets which preserves union, intersection and complementation is called an *isomorphism* or, specifically, a Boolean isomorphism. Thus the algebra of pairs (A, A') is isomorphic to the algebra of classes A.

2.92. The relationship between the pairs (A, B) and the pairs (A, A') is a many-one relation since all the pairs (A, B) for all B correspond to the single pair (A, A'). We express this relationship by writing

$$(A, B) \rightarrow (A, A')$$

and say that the set of pairs (A, B) is mapped on to the set of pairs (A, A'). To each pair (A, B) corresponds a single pair (A, A'), but each (A, A') is the 'map' of all the pairs (A, B) with first term A. This correspondence, like the isomorphism we considered before, preserves complementation, union and intersection. In fact we have

$$(A, B) \rightarrow (A, A'), (A, B)' = (A', B') \rightarrow (A', A) = (A, A')',$$
$$(A_1, B_1) \cup (A_2, B_2) = (A_1 \cup A_2, B_1 \cap B_2) \rightarrow (A_1 \cup A_2, A_1' \cap A_2')$$
$$= (A_1, A_1') \cup (A_2, A_2')$$

and

$$(A_1, B_1) \cap (A_2, B_2) = (A_1 \cap A_2, B_1 \cup B_2) \rightarrow (A_1 \cap A_2, A_1' \cup A_2')$$
$$= (A_1, A_1') \cap (A_2, A_2').$$

A many-one correspondence which preserves complementation, union and intersection is called a *homomorphism*, or specifically a Boolean homomorphism. Thus the relationship between the pairs (A, B), and the pairs (A, A') is a homomorphism—as is the relationship between the pairs (A, B) and the classes A. In the homomorphism of the pairs (A, B) on to the classes A, all the pairs $(0, B)$ map on to the zero 0, and all the pairs $(1, B)$ map on to the unit 1.

2.93. The one-to-one correspondence between $(0, B)$ and B' preserves union and intersection since

$$(0, B_1) \cup (0, B_2) = (0, B_1 \cap B_2) \leftrightarrow B_1' \cup B_2'$$

$$(0, B_1) \cap (0, B_2) = (0, B_1 \cup B_2) \leftrightarrow B_1' \cap B_2'$$

but complementation is *not* preserved, since the complement of $(0, B)$ is not one of this set of pairs.

A one-to-one correspondence which preserves union and intersection but not complementation may be called a *partial (or lattice) isomorphism*. However the mapping of the class of *all* pairs $(0, B)$ and $(1, B)$ upon B' is a Boolean homomorphism, since

$$(0, B)' = (1, B') \leftrightarrow B = (B')',$$

$$(1, B)' = (0, B') \leftrightarrow B$$

$$(0, B_1) \cup (0, B_2) = (0, B_1 \cap B_2) \leftrightarrow B_1' \cup B_2'$$

$$(0, B_1) \cup (1, B_2) = (1, B_1 \cap B_2) \leftrightarrow B_1' \cup B_2'$$

$$(1, B_1) \cup (1, B_2) = (1, B_1 \cap B_2) \leftrightarrow B_1' \cup B_2'$$

and

$$(0, B_1) \cap (0, B_2) = (0, B_1 \cup B_2) \leftrightarrow B_1' \cap B_2'$$

$$(0, B_1) \cap (1, B_2) = (0, B_1 \cup B_2) \leftrightarrow B_1' \cap B_2'$$

$$(1, B_1) \cap (1, B_2) = (1, B_1 \cup B_2) \leftrightarrow B_1' \cap B_2'.$$

2.94. The mapping of A upon $A \cup K$, for a fixed class K, preserves union and intersection because

$$A_1 \cup A_2 \rightarrow (A_1 \cup A_2) \cup K = (A_1 \cup K) \cup (A_2 \cup K)$$

and

$$A_1 \cap A_2 \rightarrow (A_1 \cap A_2) \cup K = (A_1 \cup K) \cap (A_2 \cup K)$$

but it does not preserve complementation because

$$(A \cup K)' = A' \cap K';$$

such a correspondence may be called a *partial (or lattice) homomorphism*.

2.95. A partial homomorphism (and in particular a partial isomorphism) preserves inclusion, for if $A \to A^*$ and if $A \subset B$, then $A = A \cap B$ and so

$$A^* = (A \cap B)^* = A^* \cap B^*$$

which proves that $A^* \subset B^*$.

2.96. If a Boolean homomorphism (and in particular an isomorphism) maps a Boolean algebra of elements $A, B, C, \ldots$ on to a set of elements $a, b, c, \ldots$ closed under complementation, union and intersection, then the set of elements $a, b, c, \ldots$ form a Boolean algebra.

We have to verify that the axioms 2.01–2.04 are all satisfied by the set $a, b, c, \ldots$ given that they are satisfied by the set $A, B, C, \ldots$.

We have

$$A \cup B \to a \cup b, \quad B \cup A \to b \cup a$$

and

$$A \cup B = B \cup A$$

so that

$$a \cup b = b \cup a.$$

Similarly from $A \cap B \to a \cap b$, $B \cap A \to b \cap a$ and $A \cap B = B \cap A$ follows $a \cap b = b \cap a$.

For the distributive laws we have

$$A \cup (B \cap C) \to a \cup (b \cap c),$$

$$(A \cup B) \cap (A \cup C) \to (a \cup b) \cap (a \cup c)$$

whence from $A \cup (B \cap C) = (A \cup B) \cap (A \cup C)$ follows

$$a \cup (b \cap c) = (a \cup b) \cap (a \cup c).$$

The proof of the second distributive law is obtained simply by interchanging $\cup$ and $\cap$.

If 0 is the map of 0, and 1 the map of 1 we have

$$A \cup 0 \to a \cup 0, \quad \text{and} \quad A \to a,$$

and so from $A = A \cup 0$ follows $a = a \cup 0,$

and similarly from

$$A = A \cap 1 \quad \text{follows} \quad a = a \cap 1.$$

Finally from

$$A \cup A' \to a \cup a', \quad 1 \to 1$$

and $A \cup A' = 1$

follows $a \cup a' = 1,$

and from

$$A \cap A' \to a \cap a', \quad 0 \to 0,$$

and $A \cap A' = 0$

follows $a \cap a' = 0.$

2.97. If $A \to A^*$ is a Boolean homomorphism of a Boolean algebra on to itself (so that for each element A of the algebra, A^* is an element of the algebra) and if $A \to \bar{A}$ is another Boolean homomorphism of the algebra upon itself, then the mapping

$$A \to \overline{A^*}$$

is a Boolean homomorphism.

For $(A \cup B)^* = A^* \cup B^*$

and so $\overline{(A \cup B)^*} = \overline{A^* \cup B^*} = \overline{A^*} \cup \overline{B^*}$

and similarly $\overline{(A \cap B)^*} = \overline{A^*} \cap \overline{B^*}.$

For complementation we have

$$(A')^* = (A^*)'$$

and therefore $\overline{(A')^*} = \overline{(A^*)'} = \overline{A^*}',$

which completes the proof.

EXAMPLES II

1. Prove the following relations
   .1 $A \cup B = (A \cap B) \cup (A \cap B') \cup (A' \cap B)$
   .2 $(A \cap B) \cup (C \cap D) = (A \cup C) \cap (B \cup D) \cap (B \cup C) \cap (A \cup D)$
   .3 $(A \cup B) \cap (C \cup D) = (A \cap C) \cup (B \cap D) \cup (B \cap C) \cup (A \cap D)$
   .4 $(A \cup B \cup D) \cap (A' \cup C \cup E)$
   $$= \{(A \cup B) \cap (A' \cup C)\} \cup \{(A \cup D) \cap (A' \cup E)\}$$
   .5 $\qquad\qquad A + B = (A \cap B') + (A' \cap B)$
   .6 $(A + B) \cup (B + C) = (A + C) \cup (B + C)$
   .7 $\qquad\qquad A \cup (A + B) = A \cup B$
   .8 $\qquad\qquad A + (A \cup B) = B - A$
   .9 $\qquad\qquad A \cup B = A + \{B - (A \cap B)\}$
   .91 $\qquad\qquad A + C \subset (A + B) \cup (B + C)$
   .92 $(A \cup B) \cap (A' \cup C) = (A' \cap B) \cup (A \cap C)$.

2. Establish the relations in example I, 12 and I, 13 by the decision method of § 2.74.

3. Transform to both standard forms the expressions
   $$(A \cup B)' \cap A \cap B$$
   $$(A \cup B \cup C)' \cap (A \cup B)$$
   $$(A' \cap B)' \cap (C \cup B')' \cap (C \cup A').$$

4. Prove that, for a fixed K, the mapping $A \to A + K$ is an isomorphic mapping which preserves complements.

5. Prove that the mapping
   $$(A, B) \to B',$$
   is a Boolean homomorphism.

6. If the correspondence $A \leftrightarrow a$ is an isomorphism between sets $\{A\}$, $\{a\}$ which are closed under complementation, union and intersection, and if one of these sets is a Boolean algebra, then so is the other.

7. If the correspondence $A \to a$ is a homomorphism of a set $\{A\}$ on to a set $\{a\}$ which are both closed under complementation, union and intersection, prove that if the set $\{A\}$ is a Boolean algebra then so is the set $\{a\}$.

8. Prove that the set of elements which map on a single element in a partial homomorphism is closed under union and intersection.

9. If the mappings $A \to A^*$, $A \to \bar{A}$ are partial homomorphisms of a set upon itself, prove that both the mappings
   $$A \to A^* \cup \bar{A}, \qquad A \to A^* \cap \bar{A}$$
   preserve inclusion.

10. N is a product of n distinct primes $p_1, p_2, \ldots, p_n$ and $a, b, c, \ldots$ are the various divisors of N (including unity); $a \cup b$ is the least common multiple of a, b and $a \cap b$ is their highest common factor. If 0 has the value unity and 1 the value N, and if $a' = N/a$, prove that the relations 2.01–2.04 are all satisfied by $a, b, c, \ldots$.

11. If
$$C \cup A = B, \quad C \cap A = 0$$
and
$$D \cup A = B, \quad D \cap A = 0$$
prove that
$$B \cap A = A, \quad B \cap C = C$$
and
$$C = D.$$

12. Prove that if $A \cap B = A$ the simultaneous equations
$$X \cup A = B, \quad X \cap A = 0$$
have the unique solution
$$X = A' \cap B.$$

13. Prove that the simultaneous equations
$$X \cap (A \cup B) = X, \quad A \cap (B \cup X) = A, \quad B \cap (A \cup X) = B$$
$$X \cap A \cap B = 0$$
have the unique solution
$$X = A + B.$$

CHAPTER THREE

Boolean equations

THERE are many axiom systems for Boolean algebra. In this chapter we consider an axiom system with two operations, $A \cap B$ and A' which we continue to call intersection and complementation. The axioms are:

3.1. $$A \cap B = B \cap A,$$
3.11. $$A \cap (B \cap C) = (A \cap B) \cap C.$$
3.2. If for some A, B, C
$$A \cap B' = C \cap C'$$
then
$$A \cap B = A.$$

3.21. If $$A \cap B = A$$

then $$A \cap B' = C \cap C'$$

for *any* C.

As in the previous system of axioms we require axioms to ensure that equality is preserved under intersection and complementation and so we add the axioms:

3.3. if $A = B$ then $A' = B'$, and $A \cap C = B \cap C$.

We start by remarking that each of these axioms is an axiom or a provable property of intersection and complementation in the system of the previous chapter, and so any property we derive from these axioms will certainly be provable from the previous set of axioms. It remains to show that in fact every property derivable from axioms 2.01–2.04 is derivable also from axioms 3.1–3.21, and the simplest way to prove this is to show that axioms 2.01–2.04

are themselves derivable from 3.1–3.21. To this end we first *define* union in terms of intersection by the equation

3.4. $$A \cup B = (A' \cap B')'$$

(a *provable* property in the previous chapter, and now a *definition*) and define 0, 1 by the equations:

3.41. $$0 = A \cap A', \qquad 1 = 0',$$

(for a certain A in the first instance, and for any A as we prove shortly). For inclusion we keep the definition of Chapter 2 and write

3.42. $$A \subset B \quad \text{if, and only if,} \quad A \cap B = A.$$

From the equation

$$A \cap A' = A \cap A'$$

and 3.2 it follows that

3.421. $$A \cap A = A$$

and hence, by 3.21,

$$A \cap A' = C \cap C'$$

for any C, that is

3.43. $$C \cap C' = 0$$

for any C, which proves one part of 2.04.
From 3.2 and 3.21 again it now follows that

$$A \cap B = A$$

if and only if $A \cap B' = 0$, and so $A \subset B$ is equivalent to $A \cap B' = 0$. In particular it follows that

$$A \subset A.$$

To prove that inclusion is transitive we observe that from

$$A \cap B = A \quad \text{and} \quad B \cap C = B$$

follows

$$A \cap C = (A \cap B) \cap C = A \cap (B \cap C), \qquad \text{by 3.11,}$$
$$= A \cap B = A,$$

and so

3.44. from $A \subset B$ and $B \subset C$ follows $A \subset C$.

From $A \cap A = A$ and 3.1 again we derive

3.45. $$A \cap B \subset A,$$

and from the commutative law we find that

if $$A \subset B \quad \text{and} \quad B \subset A$$

then

$$A = A \cap B = B \cap A = B$$

3.46. i.e., $A \subset B$ and $B \subset A$ entail $A = B$.

Of course if $A = B$ then $A = A \cap A = B \cap A = A \cap B$ and so $A \subset B$; similarly from $B = A$ follows $B \subset A$.

To prove that $A \cap 0 = 0$ we observe that $0 = A \cap A'$ and $A \cap A' \subset A$ (by 3.45)

and so

3.47. $$0 \subset A$$

and hence

3.48. $$A \cap 0 = 0 \cap A = 0.$$

Our next task is to prove that

3.49. $$A'' = A.$$

Since $A'' \cap A' = A' \cap A'' = 0$, therefore $A'' \subset A$; hence $A''' \subset A'$ and $A'''' \subset A''$ (taking A' and then A'' for A in $A'' \subset A$) and so by the transitivity of inclusion

$$A'''' \subset A,$$

and so

$$A'''' \cap A' = 0,$$

from which we deduce

$$A' \subset A''';$$

since we have also shown that $A''' \subset A'$ we have

$$A''' = A';$$

from this, and the fact that $A \cap A' = 0$, follows

$$A \cap A''' = 0$$

and thence in turn

$$A \subset A'',$$

$$A = A'',$$

(since $A'' \subset A$).

3.5. We can now prove the equivalence of $A \subset B$ with $B' \subset A'$. For from $A \subset B$ we have $A \cap B' = 0$ and so $B' \cap A'' = 0$, that is, $B' \subset A'$; conversely from $B' \subset A'$ follows $A'' \subset B''$, and so $A \subset B$.

It follows that

3.51. $A \subset A \cup B$

for this is equivalent to $(A \cup B)' \subset A'$, that is $A' \cap B' \subset A'$, which we have already proved.

From 3.5 we readily show that

3.52. $A \subset B$ if, and only if, $A \cup B = B$.

For $A \cup B = B$ is equivalent to $(A' \cap B')' = B$, which in turn is equivalent to $A' \cap B' = B'$, $B' \subset A'$, $A \subset B$.

3.6. The commutative law for union is an immediate consequence of axiom 3.1, since $A \cup B = (A' \cap B')' = (B' \cap A')' = B \cup A$.

To prove the associative law for union, we have

$$\begin{aligned}
A \cup (B \cup C) &= A \cup (B' \cap C')' \\
&= \{A' \cap (B' \cap C')\}' \\
&= \{(A' \cap B') \cap C'\}', \quad \text{by axiom 3.1,} \\
&= (A' \cap B')' \cup C \\
&= (A \cup B) \cup C.
\end{aligned}$$

3.61. By 3.421, with A' for A we have

$$A' \cap A' = A'$$

whence, taking complements,

$$A \cup A = A.$$

From 3.49 and the definitions of 0, 1 we have

3.62. $$A \cup A' = 1,$$

for $$A \cup A' = (A' \cap A)' = 0' = 1.$$

3.63. If $A \subset B$ then $A \cap C \subset B \cap C$ and $A \cup C \subset B \cup C$. For if $A \subset B$, then $A \cap B = A$ and so

$$A \cap B \cap C = A \cap C$$

whence
$$\begin{aligned}(A \cap C) \cap (B \cap C) &= ((A \cap C) \cap C) \cap B \\ &= A \cap (C \cap C) \cap B \\ &= A \cap B \cap C = A \cap C\end{aligned}$$

which proves that
$$A \cap C \subset B \cap C.$$

Similarly
$$(A \cup C) \cup (B \cup C) = A \cup B \cup C,$$

and so if $A \subset B$, then $A \cup B = B$ and so

$$A \cup B \cup C = B \cup C$$

whence
$$(A \cup C) \cup (B \cup C) = B \cup C$$

which proves that
$$A \cup C \subset B \cup C.$$

An immediate consequence is that from $A \subset B$ and $C \subset D$ follows $A \cup C \subset B \cup D$, for $A \cup C \subset B \cup C \subset B \cup D$; in particular from $A \subset B$, $C \subset B$ follows $A \cup C \subset B$, since $B \cup B = B$.

We prove next the distributive laws for union and intersection, starting with the particular case

3.64. $$A \cap (A \cup B) = A \cup (A \cap B).$$

By 3.51

$$A \subset A \cup B,$$

whence

$$A \cap (A \cup B) = A;$$

but $A \cap B \subset A$ so that

3.641. $$(A \cap B) \cup A = A$$

which proves that

$$A \cap (A \cup B) = (A \cap B) \cup A.$$

Since

$$A \cap (A' \cup B) \cap B' = A \cap (A \cap B')' \cap B'$$
$$= (A \cap B') \cap (A \cap B')' = 0,$$

we have, by axiom 3.2,

$$A \cap (A' \cup B) = A \cap (A' \cup B) \cap B.$$

But as we have seen $(A' \cup B) \cap B = B$
and so

3.65. $$A \cap (A' \cup B) = A \cap B.$$

From the proved relations

$$B \subset B \cup C$$

$$C \subset B \cup C$$

follow (by 3.63)

$$A \cap B \subset A \cap (B \cup C)$$

$$A \cap C \subset A \cap (B \cup C)$$

whence

3.66. $$(A \cap B) \cup (A \cap C) \subset A \cap (B \cup C).$$

Furthermore

$$A \cap (B \cup C) \cap \{(A \cap B) \cup (A \cap C)\}'$$
$$= A \cap (B \cup C) \cap \{(A' \cup B') \cap (A' \cup C')\}$$
$$= (B \cup C) \cap A \cap B' \cap (A' \cup C'), \quad \text{by 3.65}$$
$$= (B \cup C) \cap B' \cap A \cap C'$$
$$= (B \cup C) \cap (B \cup C)' \cap A$$
$$= 0 \cap A = 0, \qquad \text{by 3.48,}$$

which proves that

3.67. $\qquad A \cap (B \cup C) \subset (A \cap B) \cup (A \cap C).$

3.66, 3.67 together prove the distributive law

3.68. $\qquad A \cap (B \cup C) = (A \cap B) \cup (A \cap C).$

Taking A', B', C' for A, B, C in 3.68 and taking the complements of both sides we reach the second distributive law

3.69. $\qquad A \cup (B \cap C) = (A \cup B) \cap (A \cup C).$

By means of 3.641 we readily prove both parts of 2.03.

For $\qquad A \cup 0 = A \cup (A \cap A') = A$

and so $\qquad A' \cup 1' = A'$

whence, taking complements,

$$A \cap 1 = A.$$

We have now completed the derivations of all the relations 2.01–2.04 from the axioms 3.1–3.21 showing that these two sets of axioms have the same class of consequences, and are therefore equivalent. In the examples and in a later chapter we consider yet further equivalent axiom systems for Boolean algebra.

3.7. **Boolean equations.** If A is a given element, and X an unknown element, the equation

$$A + X = 0$$

has the unique solution $X = A$, as we have already seen (1.971), and so the equation

$$(A + X) \cup (B + Y) = 0$$

has the solutions $X = A$ and $Y = B$, with an obvious extension to more terms.

Similarly the equation

$$A \times X = 1$$

has the unique solution $X = A$ (1.98) and so the equation

$$(A \times X) \cap (B \times Y) = 1$$

has the solutions $X = A$ and $Y = B$, (with again an obvious extension to three or more terms).

The equation

3.71. $A \cup X = 1$

has the general solution $X = A' \cup U$ where U is an arbitrary element; the solution is general in the sense that this value of X satisfies the equation and for a suitable value of U, every solution has the form $A' \cup U$.

First we observe that if $X = A' \cup U$ then $A \cup X = A \cup A' \cup U = 1 \cup U = 1$. Moreover, if $A \cup X = 1$, then $A' \cap X' = 0$ and so

$$X \cup (A' \cap X') = X \cup 0 = X$$

whence $(X \cup A') \cap (X \cup X') = X$

and so $X = A' \cup X$

showing that every solution has the form $A' \cup U$ (with U having the value X).

It follows that the equation

3.72. $(A \cup X) \cap (B \cup Y) \cap (C \cup Z) = 1$

has the solution

$$X = A' \cup U \quad \text{and} \quad Y = B' \cup V \quad \text{and} \quad Z = C' \cup W$$

where U, V, W are arbitrary.

3.73. The equation

$$X \cup (A \cap Y) = K$$

has the general solution

$$X = \{U \cup (A' \cup V')\} \cap K$$
$$Y = (A' \cup K) \cap V.$$

We start by showing that these values of X and Y do satisfy the equation; in fact

$$A \cap Y = A \cap (A' \cup K) \cap V$$
$$= A \cap K \cap V$$

and so

$$X \cup (A \cap Y) = K \cap [(A \cap V) \cup U \cup (A' \cup V')]$$
$$= K \cap [U \cup (A \cap V) \cup (A \cap V)']$$
$$= K \cap 1 = K.$$

Conversely, if

$$X \cup (A \cap Y) = K$$

then $A \cap Y \subset K$ and so

$$Y \subset Y \cup A' = (A \cap Y) \cup A' \subset K \cup A'$$

so that

$$Y = (K \cup A') \cap Y$$

showing that $Y = (K \cup A') \cap V$, with the value Y of V.

Furthermore, since

$$X' \cap (A' \cup Y') = K'$$

therefore

$$X \cup (X' \cap (A' \cup Y')) = X \cup K'$$

whence

$$X \cup A' \cup Y' = X \cup K'$$

and so

$$K \cap (X \cup A' \cup Y') = K \cap X;$$

but since $X \cup (A \cap Y) = K$, therefore $X \subset K$ so that $K \cap X = X$ proving that

$$X = \{U \cup (A' \cup Y')\} \cap K$$

with the value X of U.

The equation

3.74. $X \cup Y = K$

is of course a particular case of the last equation, with $A = 1$; but the form of solution obtained by taking $A = 1$, is not symmetrical with respect to X and Y. We shall show that the general solution may be expressed in the form

$$X = K \cap (U \cup V')$$

$$Y = K \cap (U' \cup V)$$

where U, V are arbitrary. For with these values

$$X \cup Y = \{K \cap (U \cup V')\} \cup \{K \cap (U' \cup V)\}$$
$$= K \cap \{U \cup V' \cup U' \cup V\} = K \cap 1 = K,$$

and conversely, if X, Y satisfy $X \cup Y = K$ then

$$X = X \cup (Y \cap Y') = (X \cup Y') \cap (X \cup Y) = K \cap (X \cup Y'),$$

$$Y = Y \cup (X \cap X') = (X \cup Y) \cap (X' \cup Y) = K \cap (X' \cup Y),$$

so that

$$X = K \cap (U \cup V')$$

$$Y = K \cap (U' \cup V)$$

with the values X, Y of U, V.

We consider now a generalization of equation 3.73.

3.75. $X \cup (A \cap Y) \cup (B \cap Z) = K.$

The general solution is

$$X = K \cap [U \cup \{(A' \cup V') \cap (B' \cup W')\}],$$

$$Y = \{K \cup A'\} \cap V,$$

$$Z = \{K \cup B'\} \cap W.$$

We observe first that these values satisfy the equation for

$$A \cap Y = V \cap \{(A \cap K) \cup (A \cap A')\} = V \cap A \cap K,$$

$$B \cap Z = W \cap B \cap K$$

and so

$$X \cup (A \cap Y) \cup (B \cap Z)$$

$$= K \cap \{U \cup ((A' \cup V') \cap (B' \cup W')) \cup (V \cap A) \cup (W \cap B)\}$$
$$= K \cap \{U \cup ((A \cap V) \cup (B \cap W))' \cup (V \cap A) \cup (B \cap W)\}$$
$$= K \cap 1 = K.$$

Consider now any solution $X = X_1$, $Y = Y_1$, $Z = Z_1$ of the given equation; then

$$X_1 \cup (A \cap Y_1) \cup (B \cap Z_1) = K$$

so that $A \cap Y_1 \subset K$ and $B \cap Z_1 \subset K$, and therefore

$$Y_1 \subset Y_1 \cup A' = (A \cap Y_1) \cup A' \subset K \cup A'$$

whence

$$Y_1 = (K \cup A') \cap Y_1$$

and similarly

$$Z_1 = (K \cup B') \cap Z_1$$

showing that

$$Y_1 = (K \cup A') \cap V, \qquad Z_1 = (K \cup B') \cap W$$

for the values Y_1 of V, Z_1 of W.

Taking the complement of the equation satisfied by X_1, Y_1, Z_1 we find

$$X_1' \cap (A' \cup Y_1') \cap (B' \cup Z_1') = K'$$

whence

$$X_1 \cup \{(A' \cup Y_1') \cap (B' \cup Z_1')\} = X_1 \cup K'$$

and so

$$K \cap [X_1 \cup \{(A' \cup Y_1') \cap (B' \cup Z_1')\}] = K \cap X_1.$$

Since $X_1 \subset K$ therefore $K \cap X_1 = X_1$ so that

$$X_1 = K \cap [X_1 \cup \{(A' \cup Y_1') \cap (B' \cup Z_1')\}]$$

which shows that

$$X_1 = K \cap [U \cup \{(A' \cup V') \cap (B' \cup W')\}]$$

for the values X_1 of U, Y_1 of V, Z_1 of W.

There is an obvious extension of the solution to any number of classes.

The equation

3.76. $K \cap (A \cup X) = 0$

is equivalent to the two equations

$$K \cap A = 0$$

$$K \cap X = 0;$$

if the condition $K \cap A = 0$ is satisfied then the general solution of the equation is

$$X = K' \cap U$$

where U is arbitrary; for if $X = K' \cap U$ (and $K \cap A = 0$) then

$$K \cap (A \cup X) = K \cap (A \cup (K' \cap U))$$
$$= (K \cap A) \cup (K \cap K' \cap U) = 0,$$

and conversely if $K \cap X_1 = 0$ then $K' \cup X_1' = 1$

and so $X_1 = X_1 \cap 1 = X_1 \cap (X_1' \cup K') = K' \cap X_1$

showing that $X_1 = K' \cap U$ for the value X_1 of U.

The solution of the equation in two unknowns

3.77. $K \cap (A \cup X) \cap (B \cup Y) = 0$

may be deduced from that of 3.76.

For 3.77 is equivalent to the two equations

3.771. $$K \cap B \cap (A \cup X) = 0$$

3.772. $$K \cap (A \cup X) \cap Y = 0.$$

The first of these equations has a solution if, and only if,

$$A \cap B \cap K = 0$$

and, as we have seen, if this condition is satisfied the general solution is

$$X = K' \cup B' \cup U, \quad \text{where } U \text{ is arbitrary.}$$

With this value of X, the general solution of 3.772 is

$$Y = \{K \cap (A \cup X)\}' \cup V$$

where V is arbitrary.

In the same way, the solution of the equation

3.78. $$K \cap (A \cup X) \cap (B \cup Y) \cap (C \cup Z) = 0$$

in three unknowns, may now be written down. For this equation is equivalent to

$$(K \cap C) \cap (A \cup X) \cap (B \cup Y) = 0$$
$$K \cap (A \cup X) \cap (B \cup Y) \cap Z = 0;$$

the first of these is solvable if and only if

$$A \cap B \cap C \cap K = 0$$

and when this condition is satisfied we have the general solution

$$X = (K \cap C)' \cup B' \cup U = K' \cup B' \cup C' \cup U$$
$$Y = \{K \cap C \cap (A \cup X)\}' \cup V$$
$$Z = \{K \cap (A \cup X) \cap (B \cup Y)\}' \cup W,$$

where U, V, W are arbitrary.

The method clearly extends to equations in any number of unknowns.

The equation

3.79. $(A \cap X) \cup (B \cap X') = 0$

is solvable if and only if $A \cap B = 0$,
and if this condition is satisfied the general solution is
$$X = (A' \cap U) \cup (B \cap U').$$
For if $A \cap B = 0$ and $X = (A' \cap U) \cup (B \cap U')$ then

$$\begin{aligned}
A \cap X &= A \cap B \cap U', & B \cap X' &= B \cap (A \cup U') \cap (B' \cup U) \\
&= 0, & &= (B \cap U') \cap (B' \cup U) \\
& & &= (B \cap U') \cap (B \cap U')' \\
& & &= 0
\end{aligned}$$

so that 3.79 is satisfied. Conversely if X_1 is a value of X which satisfies 3.79 then

3.791. $(A \cap X_1) \cup (B \cap X_1') = 0$

whence, by complements,

$$(A' \cup X_1') \cap (B' \cup X_1) = 1$$

and so $A \cap B = A \cap B \cap 1 = (A \cap B) \cap (A' \cup X_1') \cap (B' \cup X_1)$
$$\begin{aligned}
&= (A \cap B \cap X_1') \cap (B' \cup X_1) \\
&= A \cap B \cap X_1 \cap X_1' = 0,
\end{aligned}$$

and from 3.791
$$A \cap X_1 = 0, \qquad B \cap X_1' = 0$$
whence
$$X_1 = X_1 \cup 0 = X_1 \cup (X_1' \cap B) = B \cup X_1$$
and
$$X_1 = X_1 \cap (A \cup A') = (A \cap X_1) \cup (A' \cap X_1) = A' \cap X_1$$

from which we obtain, with the value $A' \cap B' \cap X_1$ for U,

$$\begin{aligned}
(A' \cap U) \cup (B \cap U') &= (A' \cap B' \cap X_1) \cup \{B \cap (A \cup B \cup X_1')\} \\
&= (A' \cap B' \cap X_1) \cup \{B \cup (B \cap X_1')\} \\
&= (A' \cap B' \cap X_1) \cup B \\
&= (A' \cap X_1) \cup B = X_1 \cup B = X_1
\end{aligned}$$

as we wished to show.

Finally we consider the pair of simultaneous equations

3.792.
$$X \cup (Y \cap Z) = 1$$
$$Y \cup (Z \cap X) = 1.$$

The first of these equations is equivalent to the pair

$$X \cup Y = 1, \qquad X \cup Z = 1$$

and the second to

$$Y \cup Z = 1, \qquad Y \cup X = 1$$

and so the two given equations are equivalent to the three equations

$$X \cup Y = 1, \qquad Y \cup Z = 1, \qquad Z \cup X = 1.$$

The general solution of this system is

3.793. $X = U \cup V', \qquad Y = V \cup W', \qquad Z = W \cup U';$

for if X, Y, Z have these values then

$$X \cup Y = U \cup V' \cup V \cup W' = 1$$
$$Y \cup Z = V \cup W' \cup W \cup U' = 1$$
$$Z \cup X = W \cup U' \cup U \cup V' = 1$$

and conversely, if X_1, Y_1, Z_1, satisfy equations 3.792 then

$$X_1 = X_1 \cup (Y_1 \cap Y_1') = 1 \cap (X_1 \cup Y_1') = X_1 \cup Y_1'$$
$$Y_1 = Y_1 \cup Z_1'$$
$$Z_1 = Z_1 \cup X_1'$$

showing that

$$X_1 = U \cup V', \qquad Y_1 = V \cup W', \qquad Z_1 = W \cup U'$$

for the values X_1, Y_1, Z_1 of U, V, W.

3.8. Congruence. Let ρ be a relation which holds between elements of a Boolean algebra, that is to say a relation which holds for some pairs of elements, and perhaps not for others. If two elements A, B stand in the relation ρ we write $A \rho B$, and if A, B do not stand in this relationship we write $A \not\rho B$. For instance inclusion $\subset$, equality $=$, complement of, are relations between elements of a Boolean algebra.

A relation ρ is said to be an *equivalence* relation if ρ has the following three properties:

3.801. E_1: $A \rho A$ for all elements A,

3.802. E_2: if $A \rho B$ then $B \rho A$,

3.803. E_3: if $A \rho B$ and $B \rho C$ then $A \rho C$.

A relation which has property E_1 is said to be *reflexive*; one that has property E_2 is *symmetrical*, and one that has property E_3 is *transitive*. For example inclusion is reflexive and transitive, but not symmetrical; the relation "complement of" is symmetrical, because from $A = B'$ follows $B = A'$, but it is not reflexive or transitive.

An equivalence relation ρ is a *congruence* relation if ρ has (in addition to the equivalence properties E_1, E_2, E_3) the properties

3.81. C_1: if $A \rho B$ then $(A \cap C) \rho (B \cap C)$ for all C,

3.82. C_2: if $A \rho B$ then $A' \rho B'$

If ρ is a congruence relation, and $A \rho B$, then A, B are said to be congruent.

3.83. A congruence relation separates the class of elements into subclasses without common members. For let $\{A\}$ denote the class of elements congruent to A, and suppose that $\{A\}$ and $\{B\}$ have a member in common. If C is this common member then A is congruent to C, and C is congruent to B and so (by E_3) A is congruent to B, and so congruent to any member of $\{B\}$, which proves that $\{B\} \subset \{A\}$; similarly $\{A\} \subset \{B\}$ and so $\{A\} = \{B\}$. Thus either $\{A\}$, $\{B\}$ are without common members, or they coincide.

3.84. We proceed to define a Boolean algebra with elements $\{A\}$, $\{B\}$, ... We start by observing that if $C \varepsilon \{A\}$ and $D \varepsilon \{B\}$

then $C \cap D \, \varepsilon \, \{A \cap B\}$; for from $C \, \rho \, A$ follows $C \cap D \, \rho \, A \cap D$, and from $D \, \rho \, B$ follows $A \cap D \, \rho \, A \cap B$, by C_1, and hence, by E_3, $C \cap D \, \varepsilon \, \{A \cap B\}$. Similarly, if $C \, \varepsilon \, \{A\}$ then $C \, \rho \, A$ and therefore $C' \, \rho \, A'$ so that $C' \, \varepsilon \, \{A'\}$.

We define

$$\{A\} \cap \{B\} = \{A \cap B\}$$

and

$$\{A\}' = \{A'\};$$

in virtue of what we have just proved the class $\{A \cap B\}$ coincides with the class $\{C \cap D\}$ where C, D are *any* members of $\{A\}$ and $\{B\}$, and the class $\{A'\}$ coincides with the class $\{C'\}$ where C is any member of A, so that the foregoing definitions of intersection and complementation are independent of the particular representatives of the classes $\{A\}$ and $\{B\}$ used in the definition.

3.841. To prove that we have defined a Boolean algebra of classes of elements we must verify that the axioms are satisfied.

For the commutative law we have

$$\{A\} \cap \{B\} = \{A \cap B\} = \{B \cap A\} = \{B\} \cap \{A\}$$

and for the associative law

$$\{A\} \cap (\{B\} \cap \{C\}) = \{A\} \cap \{B \cap C\} = \{A \cap B \cap C\},$$

$$(\{A\} \cap \{B\}) \cap \{C\} = \{A \cap B\} \cap \{C\} = \{A \cap B \cap C\}.$$

If $\{A\} \cap \{B\}' = \{C\} \cap \{C\}'$ for some C, then

$$\{A \cap B'\} = \{C \cap C'\} = \{0\}$$

and so $A \cap B'$ is congruent to 0,
whence it follows, taking complements, that

$$A' \cup B \quad \text{is congruent to 1}$$

and therefore

$$(A' \cup B) \cap A \quad \text{is congruent to } 1 \cap A$$

whence

$$A \cap B \quad \text{is congruent to } A$$

and so

$$\{A \cap B\} = \{A\}$$

from which we conclude, as required, that

$$\{A\} \cap \{B\} = \{A\}.$$

    Conversely, if

$$\{A\} \cap \{B\} = \{A\}$$

then

$$\{A \cap B\} = \{A\}$$

that is, A is congruent to $A \cap B$
and so $A \cap B'$ is congruent to

$$A \cap B \cap B' = A \cap 0 = 0 = C \cap C'$$

i.e.

$$\{A \cap B'\} = \{C \cap C'\}$$

and so

$$\{A\} \cap \{B\}' = \{C\} \cap \{C\}'$$

which completes the proof.

3.85. The mapping of A upon $\{A\}$ is a homomorphism, for

$$A \cap B \rightarrow \{A \cap B\} = \{A\} \cap \{B\}$$

and

$$A' \rightarrow \{A'\} = \{A\}'$$

3.86. *Conversely a homomorphism of a Boolean algebra $\mathscr{A}$ upon a Boolean algebra $\mathscr{A}^*$ determines a congruence relation for the elements of $\mathscr{A}$*; we define the relation ρ by the condition that, for any two elements A, B of $\mathscr{A}$, $A \rho B$ if and only if A and B map on to the *same* element of $\mathscr{A}^*$ and proceed to show that ρ is a congruence relation. Certainly ρ is an equivalence relation, and it remains only to show that if A, B map on to the same element of $\mathscr{A}^*$ then so do $A \cap C$ and $B \cap C$ for any C and, further, that A', B' map on to the same elements of $\mathscr{A}^*$.

Let A^* be the map of A in $\mathscr{A}^*$, and C^* that of C, then

$$A \cap C \rightarrow A^* \cap C^*, \qquad B \cap C \rightarrow A^* \cap C^*$$

so that

$$(A \cap C) \rho (B \cap C),$$

and
$$A' \to A^{*'}, \qquad B' \to A^{*'}$$
so that
$$A' \rho B',$$

which completes the proof that this relation ρ is a congruence relation.

3.87. *If $\{A\}$ is the class of elements which map upon the same element A^* of $\mathscr{A}^*$ as A,* (and so $\{A\}$ is the class of elements which bears the relation ρ to A) *then the relationship between $\{A\}$ and A^* is an isomorphism.* For A^* is uniquely determined as the map of *any* member of the class $\{A\}$, and conversely, given A^*, the class $\{A\}$ is uniquely determined as the class of all elements of $\mathscr{A}$ which map upon A^*, and

$$\{A\} \cap \{B\} = \{A \cap B\} \leftrightarrow A^* \cap B^*$$
$$\{A\}' = \{A'\} \leftrightarrow A^{*'}.$$

3.88. *If $\{U\}$, $\{V\}$ are the classes of elements of $\mathscr{A}$ which map upon the zero 0^* and unit 1^* of the Boolean algebra $\mathscr{A}^*$ in a homomorphism of $\mathscr{A}$ on to $\mathscr{A}^*$ then $\{U\}$, $\{V\}$ contain the zero 0 and unit 1 respectively of $\mathscr{A}$.*

For $V = V \cap 1$ and so if E^* is the map of 1 then
$$V \cap 1 \to 1^* \cap E^*, \qquad V \to 1^*$$
so that
$$E^* = 1^* \cap E^* = 1^*$$

showing that 1 is mapped upon 1^*.
It follows that
$$1' \to 1^{*'}$$
i.e.
$$0 \to 0^*.$$

3.89. *If $\{U\}$, $\{V\}$ are the classes of elements of $\mathscr{A}$ which map upon the zero and unit, 0^*, 1^* of $\mathscr{A}^*$, in a homomorphic mapping of $\mathscr{A}$ on to $\mathscr{A}^*$, and if A maps upon A^*, then the class $\{A\}$ is the class of elements formed by taking the symmetric difference of A and any member of $\{U\}$, and equally the class formed from the cross of A with any member of $\{V\}$.*

We observe first that if $A \to A^*$ and $B \to B^*$ then $A + B \to A^* + B^*$, and $A \times B \to A^* \times B^*$ for

$$A + B = (A \cup B) \cap (A' \cup B')$$
$$\to (A^* \cup B^*) \cap (A^{*'} \cup B^{*'}) = A^* + B^*,$$

and consequently

$$A \times B \to A^* \times B^*.$$

If K is any member of $\{U\}$, so that K maps upon 0^*, then

$$A + K \to A^* + 0^* = A^*$$

so that $A + K$ belongs to $\{A\}$; conversely if B is any member of the class $\{A\}$, then since

$$B = (A + A) + B = A + (A + B)$$

and

$$A + B \to A^* + A^* = 0^*,$$

so that $A + B$ belongs to $\{U\}$, therefore B is the symmetric difference of A and a member of $\{U\}$.

Similarly if L is any member of $\{V\}$, so that L maps upon 1^*, then

$$A \times L \to A^* \times 1^* = A^*$$

so that $A \times L$ belongs to $\{A\}$; conversely if B is any member of $\{A\}$ then, since

$$B = (A \times A) \times B = A \times (A \times B)$$

and

$$A \times B \to A^* \times A^* = 1^*,$$

therefore $A \times B$ belongs to $\{V\}$ and B is the cross of A and a member of $\{V\}$.

3.9. We conclude this chapter by proving the independence of axioms 3.1, 3.11, 3.2, 3.21. Independence proofs for the first three axioms are easy, but the independence of the fourth axiom is rather more difficult to establish.

3.91. To exhibit the independence of axiom 3.1 we consider an algebra of three elements 0, 1, a and relations of intersection and complementation defined by the tables

$\cap$	0	1	a
0	0	0	0
1	0	1	1
a	0	a	a

$0' = 1$
$1' = 0$
$a' = 0$

Since $1 \cap a = 1$, but $a \cap 1 = a$ it follows that 3.1 is not satisfied. The verification of B_2 requires the examination of 27 equations like

$$0 \cap (1 \cap a) = (0 \cap 1) \cap a$$

and we shall take this verification for granted.
For any element x (where x is 0, 1 or a) we find

$$x \cap x' = 0$$

and

$$x \cap y = 0$$

only when one of x, y is 0; thus if $x \cap y' = 0$ we have $x = 0$ or $y = 1, a$; if $x = 0$, $x \cap y = x$; if $y = 1$, $x \cap 1 = x$; if $y = a$, $x \cap a = x$ showing that 3.2 is satisfied. Moreover if $x \cap y = x$ then either $x = 0$ in which case $x \cap y' = 0$; or $x = 1, y = 1$ or a, so that $y' = 0$ and $x \cap y' = 0$; or $x = a$, $y = 1$ or a, and again $y' = 0$, showing that 3.21 is satisfied.
3.92. In a similar way, by means of the tables

$\cap$	a	b	c
a	a	c	b
b	c	b	a
c	b	a	c

$a' = a$
$b' = c$
$c' = b$

we may show the independence of 3.11. We observe first that the table is symmetrical so that 3.1 is satisfied.

But
$$0 \cap (b \cap c) = 0 \cap 0 = 0$$
$$(0 \cap b) \cap c = c \cap c = c$$

showing that 3.11 is broken.

We have $a \cap b = c, \quad a \cap c = b, \quad b \cap c = a$

and $a \cap a = a, \quad b \cap b = b, \quad c \cap c = c$

and $a \cap a' = b \cap b' = c \cap c' = a$

from which it readily follows that 3.2 and 3.21 are satisfied. 3.93. For axiom 3.2 we use the table

$\cap$	0	1
0	0	0
1	0	1

$0' = 0$
$1' = 0$

Since the table for intersection is simply a multiplication table it is clear that 3.1, 3.11 are satisfied. Clearly $0 \cap 0' = 1 \cap 1' = 0$. Moreover $x \cap y = x$ if $x = 0$, or $x = y = 1$, and in these cases $x \cap y' = 0$, showing that 3.21 is satisfied. But 3.2 is not, since

$$1 \cap 0' = 0 \cap 0'$$

but

$$1 \cap 0 = 0.$$

3.94. To exhibit the independence of the fourth axiom we consider an algebra of classes of numbers. We denote by $[a]$ the class of all numbers $a, a + 1, a + 2, \ldots$ which are greater than or equal to a. In the algebra we are setting up every class A is the union of a class $[a]$ and a finite class $\sigma(A)$ of numbers *less than* $a - 1$ (if $a = 1$ or 2, $\sigma(A)$ is null), and we take the totality of such classes as elements of the algebra. The unit is the class of all numbers

1, 2, 3, ...; and the zero is the empty class. We denote by $\tau(A)$ the class of numbers less than or equal to $a - 1$ which are not members of $\sigma(A)$, and define the complement of A to be $\tau(A) \cup [a + 1]$. For instance if

$$A = 1, 3, 6, 7, 8, \ldots$$

then $\sigma(A) = 1, 3$ and $\tau(A) = 2, 4, 5$ and $A' = 2, 4, 5, 7, 8, 9, \ldots$. The intersection of two elements A and B is simply the common part of the classes A, B which is necessarily a member of the same set of classes.

For instance if

$$A = 1, 3, 6, 7, 8, 9 \ldots$$

$$B = 3, 5, 6, 11, 12, 13, \ldots$$

then

$$A \cap B = 3, 6, 11, 12, 13, \ldots .$$

Of course since

$$A = \sigma(A) \cup [a]$$

and

$$B = \sigma(B) \cup [b]$$

we have

$$A \cap B = \{\sigma(A) \cap \sigma(B)\} \cup \{\sigma(A) \cap [b]\} \cup \{\sigma(B) \cap [a]\} \cup \{[a] \cap [b]\};$$

since $\sigma(A)$ and $\sigma(B)$ are finite classes, so are

$$\sigma(A) \cup \sigma(B), \qquad \sigma(A) \cap [b], \qquad \sigma(B) \cap [a]$$

and none of their members exceed the greater of $a - 2$, $b - 2$ whereas $[a] \cap [b]$ consists of all numbers greater than or equal to the greater of a and b, showing that $A \cap B$ is an element of the algebra.

Since intersection is just class intersection, axioms 3.1, 3.11 are automatically satisfied. The example

$$A = 1, \quad 4, 5, 6, \ldots, \qquad B = 1, 2, \quad 4, 5, 6, \ldots$$

$$A \cap B = A, \qquad\qquad B' = \quad 3, \quad 5, 6, \ldots$$

$$A \cap B' = 5, 6, 7, \ldots$$

and $\quad C = 1, \quad 3, 4, 5, 6, \dots, \qquad C' = 2, \quad 4, 5, 6, \dots$

$$C \cap C' = 4, 5, 6, \dots$$

shows that axiom 3.21 is *not* satisfied. (The fact that $C \cap C'$ is *not* the same for all classes C is enough to show that 3.21 is not satisfied.)

It remains to verify that axiom 3.2 is satisfied.

We observe first that for any class $C = \sigma(C) \cup [c]$, since $\sigma(C) \cap \tau(C) = 0$, $\sigma(C) \cap [c + 1] = 0$ and $\tau(C) \cap [c] = 0$, therefore

$$C \cap C' = [c + 1].$$

We consider what relationship there must be between a, b in order that

$$A \cap B' = C \cap C'.$$

We start by showing that if $a \geqslant b$ and

$$A \cap B' = C \cap C'$$

for some C, then $\sigma(A) \subset B$.

For let $A \cap B' = [c + 1]$.

Since B' does not contain b, and $A \cap B'$ is a run of numbers $x \geqslant c + 1$, therefore $A \cap B'$ contains no number less than b; thus if $\sigma(A)$ contains a number less than b it belongs to $\sigma(B)$, as it is not in B'. Moreover A does not contain $a - 1$ so that $A \cap B'$ does not contain $a - 1$, and therefore does not contain any number less than a, accordingly no member of $\sigma(A)$ belongs to B', and so every number in $\sigma(A)$ belongs to B. Thus if $a \geqslant b$ and $A \cap B' = C \cap C'$ then

$$\sigma(A) \subset B \quad \text{and} \quad [a] \subset [b] \subset B \quad \text{so that}$$

$$A = \sigma(A) \cup [a] \subset B$$

and therefore

$$A \cap B = A.$$

But if $a < b$, then $b - 1$ belongs to both B' and A and so to $A \cap B'$, whereas b itself does not belong to $A \cap B'$, proving that $A \cap B'$ is not a run of numbers $c, c + 1, c + 2, \ldots$ and so not an intersection $C \cap C'$, for any C. This completes the proof that if

$$A \cap B' = C \cap C'$$

for some C, then $A \cap B = A$.

EXAMPLES III

Draw the consequences listed in examples 8–47, from axioms A1–A7.
A1. A set K with elements $A, B, C, \ldots$ is closed under union, and if $A = B$ then $A \cup C = B \cup C$.
A2. K is closed under complementation, and if $A = B$ then $A' = B'$.
A3. $A \cup B = B \cup A$.
A4. $(A \cup B) \cup C = A \cup (B \cup C)$.
A5. $A \cup A = A$.
A6. $(A' \cup B')' \cup (A' \cup B)' = A$.
A7. $A \cap B = (A' \cup B')'$.

8. $(A \cap B) \cup (A \cap B') = A$.
9. $A \cup A' = A' \cup A''$.
10. $A'' = A$.
11. $A \cup A' = B \cup B'$.
12. Defining $1 = A \cup A'$ prove that 1 is unique and that $1 = A' \cup A$.
13. Defining $0 = 1'$, prove that 0 is unique.
14. If $A' = B'$ then $A = B$.
15. $0 \cup A = A$.
16. $A \cap 1 = A$.
17. $A \cap A' = 0$.
18. $A \cap B = B \cap A$.
19. $(A \cap B) \cap C = A \cap (B \cap C)$.
20. $A \cup B = (A' \cap B')'$.
21. $A \cap A = A$.
22. $A \cup 1 = 1$.
23. $A \cap 0 = 0$.

24. $A \cup (A \cap B) = A$.
25. $A \cap (A \cup B) = A$.
26. If $A' \cup B = B' \cup A = 1$, then $A = B$.
27. If $A \cup B = 1$ and $A \cap B = 0$ then $A' = B$.
28. The union of terms obtained from $A \cap B \cap C$ by replacing none, one, two or three terms by their complements equals 1.
29. If U, V are any two terms obtained from $A \cap B \cap C$ as in example 28, then $U \cap V = 0$.
30. $(A \cap B) \cup (A \cap C) = (A \cap B \cap C) \cup (A \cap B \cap C') \cup (A \cap B' \cap C)$.
31. $[A \cap (B \cup C)]' = (A \cap B' \cap C') \cup (A' \cap B \cap C) \cup (A' \cap B \cap C')$
$$\cup (A' \cap B' \cap C) \cup (A' \cap B' \cap C').$$
32. $(A \cap B) \cup (A \cap C) \cup [A \cap (B \cup C)]' = 1$.
33. $[(A \cap B) \cup (A \cap C)] \cap [A \cap (B \cup C)]' = 0$.
34. $A \cap (B \cup C) = (A \cap B) \cup (A \cap C)$.
35. $A \cup (B \cap C) = (A \cup B) \cap (A \cup C)$.
36. If $A \cup B = B$ then $A \cap B = A$, and conversely.
37. If $A \cup B = B$ then $A' \cup B = 1$, and conversely.
38. If $A \cup B = B$ then $A \cap B' = 0$, and conversely.
39. We define $A \subset B$ to be equivalent to $A \cap B' = 0$; give three other equivalent forms.
40. If union and complementation satisfy the table

$\cup$	0	1	2	3	4	5	$'$
0	0	1	2	3	4	5	1
1	1	1	1	1	1	1	0
2	2	1	2	1	1	2	3
3	3	1	1	3	3	1	2
4	4	1	1	4	4	1	5
5	5	1	5	1	1	5	4

show that all the axioms except A3 are satisfied by the elements 0, 1, 2, 3, 4, 5.
41. Show that the system given by the table

$\cup$	0	1	2	3	$'$
0	0	1	2	3	1
1	1	1	2	0	0
2	2	2	2	1	3
3	3	0	1	0	2

satisfies all the axioms except A4.

42. Show that the system given by the table

∪	0	1	2	′
0	0	1	2	1
1	1	1	1	0
2	2	1	1	2

satisfies all the axioms except A5.

43. Show that the system given by the table

∪	0	1	2	3	4	5	′
0	0	1	2	3	4	5	1
1	1	1	1	1	1	1	0
2	2	1	2	1	1	1	3
3	3	1	1	3	1	1	2
4	4	1	1	1	4	1	5
5	5	1	1	1	1	5	4

satisfies all the axioms except A6.

44. Prove that the general solution of the equation
$$X + Y = A + B$$
is $X = A + U$, $Y = B + U$, where U is arbitrary.

45. Find the general solution of the equations
   .1 $X \times Y = A \times B$,
   .2 $C \cap X = C \cap Y$,
   .3 $A \cap X = 0$.

46. Show that the equation
$$A \cup X = B$$
has a solution if and only if $A \subset B$, and if this condition is satisfied show that the general solution is
$$X = (U \cup A') \cap B.$$

47. Find the condition for the equation
$$A \cap X = B$$
to have a solution, and when this condition is satisfied, find the general solution.

CHAPTER FOUR

Sentence logic

4. We start this chapter by showing that Boolean algebra is also the *algebra of sentences*. If A, B are classes then we know that

$$\text{``}x \, \varepsilon \, A \cup B\text{''} \quad \text{if and only if} \quad \text{``}x \, \varepsilon \, A \quad \text{or} \quad x \, \varepsilon \, B\text{''}$$

$$\text{``}x \, \varepsilon \, A \cap B\text{''} \quad \text{if and only if} \quad \text{``}x \, \varepsilon \, A \quad \text{and} \quad x \, \varepsilon \, B\text{''}.$$

$$\text{``}x \, \varepsilon \, A' \, \text{''} \qquad \text{if and only if} \quad \text{``}x \notin A\text{''}$$

(i.e. x does not belong to A).

Thus if we let A, B stand for the sentences $x \, \varepsilon \, A$, $x \, \varepsilon \, B$ then "$A \cup B$" is equivalent to "A or B", "$A \cap B$" is equivalent to "A and B", and "A'" to "not A". Since $x \, \varepsilon \, 0$ is false for every x, 0 stands for a universally false sentence, and conversely, since $x \, \varepsilon \, 1$ is true for every x, 1 stands for a universally true sentence. A proved formula like "$A \cup A' = 1$", now reads "A or not A is universally true", and "$A \cap A' = 0$" reads "A and not A is universally false".

We may find a use also for the inclusion sign in this interpretation. Since $A \subset B$ if and only if $A' \cup B = 1$ we appropriate $A \subset B$ to stand for the sentence "not A or B"; but $A \subset B$ if and only if $x \, \varepsilon \, A$ implies $x \, \varepsilon \, B$, and so in fact we are committing ourselves to write "$A \subset B$" for "A implies B". Thus "A implies B" if and only if "B or not A".

4.01. The result of paragraph 2.74 may now be expressed by saying that a sentence (built up sentence variables by means of "and", "or" and "not") is universally true if and only if it is true for every replacement of the variables by true or false sentences 1, 0.

4.02. The tabular method of testing for provability given in § 2.74 is called the method of *truth tables* in sentence logic. Thus for instance to test whether

$$A \cap B \cap (A \cup B)' = 0$$

we set up the table

A	B	$A \cup B$	$(A \cup B)'$	$A \cap B$	$(A \cap B) \cap (A \cup B)'$
0	0	0	1	0	0
0	1	1	0	0	0
1	0	1	0	0	0
1	1	1	0	1	0

In the first two columns we enter the four pairs of values of A, B and then in succeeding columns, in each row we enter the value of the expression at the head of the column for the values of A, B in the appropriate row. The final column shows that the sentence $(A \cap B) \subset (A \cup B)$ is valid. As another example we consider the sentence

$$(A' \cup B)' \cup (A \cup C)' \cup (B \cup C) = 1.$$

The table now contains $2^3 = 8$ rows, since there are 8 trios of values for A, B, C.

A	B	C	A'	$A' \cup B$	$(A' \cup B)'$	$(A \cup C)'$	$(B \cup C)$	
0	0	0	1	1	0	1	0	1
0	0	1	1	1	0	0	1	1
0	1	0	1	1	0	1	1	1
0	1	1	1	1	0	0	1	1
1	0	0	0	0	1	0	0	1
1	0	1	0	0	1	0	1	1
1	1	0	0	1	0	0	1	1
1	1	1	0	1	0	0	1	1

In the final column we enter the values of

$$(A' \cup B)' \cup (A \cup C)' \cup (B \cup C)$$

which are found to be 1 in every row proving that "(A implies B) implies [(C or A) implies (C or B)]" is universally true.

4.1. There is of course no interest in simply repeating the work of the previous chapters, reading union, intersection and complementation as disjunction (or), conjunction (and) and negation (not). We may however give a rather different account of sentence logic, in which the elements 0, 1 play no part and we derive the class of universally true sentences from certain initial sentences and certain transformation rules (called rules of inference.)

4.2. To emphasize the new viewpoint we now use small *italic* letters for sentence variables, and the signs $\vee$, &, $\neg$, $\rightarrow$ (read or, and, not, implies) in place of $\cup$, $\cap$, $'$, $\subset$.

For initial sentences (axioms) we take

4.21. $$(p \vee p) \rightarrow p$$

4.22. $$p \rightarrow (p \vee q)$$

4.23. $$(p \vee q) \rightarrow (q \vee p)$$

4.24. $$(p \rightarrow q) \rightarrow [(r \vee p) \rightarrow (r \vee q)]$$

where, however, implication does not appear in its own right, and $p \rightarrow q$ is to be taken as a mere abbreviation for $\neg p \vee q$. Although we read "$\vee$", "$\neg$" as "or", "not" it is not intended that the signs "$\vee$", "$\neg$" have meanings assigned to them, except in so far as the axioms themselves may be considered to give the signs a meaning. Sentences are built up from sentence variables; thus p is a sentence, and if P, Q are sentences then so are $\neg P, P \vee Q$. Note the use of capital letters in this context; P, Q are not themselves sentences, but stand for sentences. We shall maintain this convention throughout the Chapter. To construct further provable sentences from the axioms we have two transformation, or inference rules:

If a proved sentence contains a variable p, then the sentence obtained by replacing p by a sentence, is a proved sentence. We call this rule *substitution.*

If P, and P → Q, are proved sentences, then so is Q.

This rule may be called *detachment*.

For example, writing $p \lor \neg q$ for q in the axiom 4.22 we obtain the proved sentence

$$p \to \{p \lor (p \lor \neg q)\}.$$

4.3. We observe first that, writing $\cup$ for $\lor$ and $'$ for $\neg$, all the axioms take only the value 1, when we substitute 0, 1 for the variables in all possible ways. Thus

4.21. becomes $(p \cup p)' \cup p = p' \cup p = 1$;

4.22. becomes $p' \cup (p \cup q) = 1 \cup q = 1$;

4.23. becomes $(p \cup q)' \cup (q \cup p) = (p \cup q)' \cup (p \cup q) = 1$; and

4.24. becomes $(p' \cup q)' \cup (r \cup p)' \cup (r \cup q)$

which has the value 1 as we have already shown.

Thus the axioms take only the value 1. The property of taking a constant value is of course preserved under substitution, for if some sentence which contains a variable p, takes only the value 1, whether p is given the value 0 or 1, then necessarily the sentence obtained by replacing p by some sentence P still takes only the value 1. Finally we observe that if P and $P \to Q$ both equal 1, then Q equals 1, for $1 \to 0 = 1' \cup 0 = 0 \cup 0 = 0$.

We have shown that all sentences derived from the axioms by the rules of inference take only the value 1. Conversely we may show that every Boolean expression which takes only the value 1 (i.e. every universally true sentence) may be derived from axioms 4.21–4.24 so that these axioms are *complete* with respect to the truth tables. We give the argument first in outline and then fill in the details.

The essential steps of the proof are these: we show that $P \& Q$ is provable if and only if both P and Q are provable; that if P is provable then so is $R \lor P \lor S$; that $P \lor \neg P$ is provable, that a sentence is provable if and only if we can prove its standard form

$$D_1 \& D_2 \& \dots D_k$$

where each D_i is a disjunction of variables or negated variables. We may then show that a sentence S which is universally true has a universally true standard form which necessarily contains some pair $p \lor \neg p$ in each of its disjunctions D_i from which it follows that each D_i is provable, their conjunction is provable, and so finally S is provable.

4.4. From axiom 4.24, substituting $\neg r$ for r, we obtain

$$(p \to q) \to [(r \to p) \to (r \to q)]$$

and from this, by substitution again,

4.41. $$(Q \to R) \to [(P \to Q) \to (P \to R)].$$

It follows that from $P \to Q$ and $Q \to R$ we may derive $P \to R$, for from $Q \to R$ and 4.41 we infer, by detachment, that

$$(P \to Q) \to (P \to R)$$

and hence from $P \to Q$, by detachment again we infer $P \to R$.

4.42. To express the fact that a sentence Q may be inferred from sentences $P_1, P_2, \dots , P_n$ we shall write

$$P_1, P_2, \dots , P_n \vdash Q.$$

If Q may be inferred from the axioms alone, i.e. if Q is provable, we write $\vdash Q$.

Thus we have just proved

4.43. $$P \to Q, Q \to R \vdash P \to R.$$

From 4.23, by substitution, we have

$$(P \lor Q) \to (Q \lor P)$$

whence, by detachment, from $P \lor Q$ we may infer $Q \lor P$, that is

4.44. $$(P \lor Q) \vdash (Q \lor P).$$

Since $Q \to (Q \lor \neg P) \to (\neg P \lor Q)$ by .22, .23, we have

4.441. $$Q \vdash (P \to Q).$$

Again from 4.24, and detachment we prove

4.45. $P \to Q \vdash (R \lor P) \to (R \lor Q)$.

Then, from 4.44, 4.45 we derive from $P \to Q$,

$(P \lor R) \to (R \lor P), (R \lor P) \to (R \lor Q), (R \lor Q) \to (Q \lor R)$

and hence, by 4.43,
$$(P \lor R) \to (Q \lor R),$$
and so we have proved

4.46. $P \to Q \vdash (P \lor R) \to (Q \lor R)$.

Similarly we may show that

4.461. $P \to Q, R \to S \vdash P \lor R \to Q \lor S$

for from $P \to Q$ follows $P \lor R \to Q \lor R$, and from $R \to S$ follows $Q \lor R \to Q \lor S$.

We readily prove now that

4.47. $p \to p$.

For from 4.22, $p \to (p \lor p)$

and from 4.21, $(p \lor p) \to p$,

whence by 4.43,
$$p \to p,$$
i.e.
$$\neg p \lor p,$$
and so from 4.44,
$$p \lor \neg p.$$

By substituting $\neg p$ for p we find
$$\neg \neg p \lor \neg p$$
and thence, by 4.44,
$$\neg p \lor \neg \neg p$$
i.e.

4.48. $p \to \neg \neg p$.

Again by substituting $\neg p$ for p,

$$\neg p \to \neg\neg\neg p$$

and from axiom 4.24

$$(\neg p \to \neg\neg\neg p) \to [(p \vee \neg p) \to (p \vee \neg\neg\neg p)]$$

whence by detachment

$$(p \vee \neg p) \to (p \vee \neg\neg\neg p)$$

and, again by detachment, from 4.47,

$$p \vee \neg\neg\neg p$$

whence

$$\neg\neg\neg p \vee p$$

i.e.

$$\neg\neg p \to p.$$

Thus we have proved both $p \to \neg\neg p$ and $\neg\neg p \to p$.
If two sentences P, Q are such that $P \to Q$, $Q \to P$ are both provable, we say that P, Q are equivalent and write

$$P \leftrightarrow Q.$$

If $P \leftrightarrow Q$ then of course $Q \leftrightarrow P$ since each stands for $P \to Q$ and $Q \to P$. If $P \leftrightarrow Q$ and P is provable, then so is Q, for Q follows from P and $P \to Q$.
In particular, then,

4.49. $$p \leftrightarrow \neg\neg p.$$

4.5. It follows immediately from 4.45 that

$$P \leftrightarrow Q \vdash (R \vee P) \leftrightarrow (R \vee Q),$$

from 4.46 that

$$P \leftrightarrow Q \vdash (P \vee R) \leftrightarrow (Q \vee R),$$

and from 4.461 that

$$P \leftrightarrow Q, R \leftrightarrow S \vdash P \vee R \leftrightarrow Q \vee S.$$

We prove next that

4.51. $(p \to q) \to (\neg q \to \neg p).$

For, from

$$q \to \neg\neg q$$

and 4.45 we infer

$$(\neg p \lor q) \to (\neg p \lor \neg\neg q)$$

and hence by 4.44

$$(\neg p \lor q) \to (\neg\neg\neg q \lor \neg p)$$

i.e.

$$(p \to q) \to (\neg q \to \neg p).$$

It follows that

4.52. $(P \to Q) \vdash (\neg Q \to \neg P),$

and

4.53. $(P \leftrightarrow Q) \vdash (\neg P \leftrightarrow \neg Q).$

If $\mathscr{S}(P)$ is any sentence which contains P, and if $P \leftrightarrow Q$ then

$$\mathscr{S}(P) \leftrightarrow \mathscr{S}(Q).$$

For by 4.5, 4.53 equivalence is preserved under disjunction and negation, and so equivalence is preserved as we build $\mathscr{S}(P)$ from P alongside $\mathscr{S}(Q)$ from Q.

For example, consider the sentence

$$\neg((R \lor P) \lor S);$$

we have

$$P \leftrightarrow Q$$

$$R \lor P \leftrightarrow R \lor Q$$

$$(R \lor P) \lor S \leftrightarrow (R \lor Q) \lor S$$

and finally

$$\neg\{(R \lor P) \lor S\} \leftrightarrow \neg\{(R \lor Q) \lor S\}.$$

As another example we take the sentence

$$(P \lor R) \lor \neg(S \lor \neg P).$$

We build up the two parts of this sentence simultaneously

$$P \leftrightarrow Q \qquad\qquad\qquad P \leftrightarrow Q$$

$$P \vee R \leftrightarrow Q \vee R \qquad\qquad \neg P \leftrightarrow \neg Q$$

$$S \vee \neg P \leftrightarrow S \vee \neg Q$$

$$\neg \{S \vee \neg P\} \leftrightarrow \neg \{S \vee \neg Q\}$$

whence, by 4.5,

$$(P \vee R) \vee \neg \{S \vee \neg P\} \leftrightarrow (Q \vee R) \vee \neg \{S \vee \neg Q\}.$$

In particular, from 4.49, we have

$$\mathscr{S}(p) \leftrightarrow \mathscr{S}(\neg\neg p).$$

4.6. We introduce

$$P \,\&\, Q$$

as an abbreviation for $\neg(\neg P \vee \neg Q)$.

Since $\qquad\qquad \neg P \vee \neg Q \leftrightarrow \neg Q \vee \neg P$

therefore $\qquad \neg(\neg P \vee \neg Q) \leftrightarrow \neg(\neg Q \vee \neg P)$

that is,

$$P \,\&\, Q \leftrightarrow Q \,\&\, P.$$

4.61. $\qquad\qquad A \rightarrow P, B \rightarrow P \vdash (A \vee B) \rightarrow P$

For $\qquad\qquad A \rightarrow P \vdash A \vee B \rightarrow P \vee B,$

$$B \rightarrow P \vdash P \vee B \rightarrow P \vee P,$$

and since (by 4.21) $P \vee P \rightarrow P$, we have 4.61.

4.62. $\qquad\qquad P \rightarrow A, P \rightarrow B \vdash P \rightarrow (A \,\&\, B).$

For $\qquad\qquad P \rightarrow A \vdash \neg A \rightarrow \neg P$

$$P \rightarrow B \vdash \neg B \rightarrow \neg P$$

and from 4.61

$$\neg A \rightarrow \neg P, \neg B \rightarrow \neg P \vdash (\neg A \vee \neg B) \rightarrow \neg P$$

and

$$(\neg A \vee \neg B) \rightarrow \neg P \leftrightarrow P \rightarrow (A \,\&\, B).$$

4.63. $P \vee (Q \vee R) \leftrightarrow Q \vee (P \vee R)$

For $R \rightarrow R \vee P \rightarrow P \vee R$
and so $Q \vee R \rightarrow Q \vee (P \vee R)$;
moreover $P \rightarrow P \vee R \rightarrow (P \vee R) \vee Q \rightarrow Q \vee (P \vee R)$
and so (by 4.61)

4.631. $P \vee (Q \vee R) \rightarrow Q \vee (P \vee R)$;

interchanging P, Q it follows that

4.632. $Q \vee (P \vee R) \rightarrow P \vee (Q \vee R)$;

from .631, .632 we obtain 4.63.

From 4.63 we derive the associative law for disjunction; for we
have $Q \vee R \rightarrow R \vee Q$ and so

$$P \vee (Q \vee R) \rightarrow P \vee (R \vee Q)$$

and

$$P \vee (R \vee Q) \rightarrow R \vee (P \vee Q) \rightarrow (P \vee Q) \vee R$$

which prove that

4.64. $P \vee (Q \vee R) \rightarrow (P \vee Q) \vee R$.

Hence

$R \vee (Q \vee P) \rightarrow (R \vee Q) \vee P \rightarrow (Q \vee R) \vee P \rightarrow P \vee (Q \vee R)$;

but $(P \vee Q) \vee R \rightarrow (Q \vee P) \vee R \rightarrow R \vee (Q \vee P)$
and so

$$(P \vee Q) \vee R \rightarrow P \vee (Q \vee R)$$

whence
4.65. $P \vee (Q \vee R) \leftrightarrow (P \vee Q) \vee R$.

Taking $\neg P$, $\neg Q$, $\neg R$ for P, Q, R in .65 we find

$$\neg P \vee (\neg Q \vee \neg R) \leftrightarrow (\neg P \vee \neg Q) \vee \neg R$$

and negating both sides of this equivalence we obtain

4.66. $P \mathbin{\&} (Q \mathbin{\&} R) \leftrightarrow (P \mathbin{\&} Q) \mathbin{\&} R$

which is the associative law for conjunction.

Since, by 4.65,

$$\neg P \vee (\neg Q \vee (P \,\&\, Q)) \leftrightarrow (\neg P \vee \neg Q) \vee (P \,\&\, Q)$$

and since $P \,\&\, Q$ stands for $\neg(\neg P \vee \neg Q)$ so that $(\neg P \vee \neg Q) \vee (P \,\&\, Q)$ is derivable from $p \vee \neg p$, therefore

$$\neg P \vee (\neg Q \vee (P \,\&\, Q))$$

is provable, that is

4.67. $P \to (Q \to (P \,\&\, Q))$.

is provable.

It follows that

4.671. $P, Q \vdash P \,\&\, Q$.

Conversely $(P \,\&\, Q) \to P$, for $(\neg P \vee \neg Q) \vee P \leftrightarrow (\neg Q \vee \neg P) \vee P \leftrightarrow \neg Q \vee (\neg P \vee P)$ and $\neg Q \vee (\neg P \vee P)$ follows from 4.47 and axioms 4.22, 4.23.

The following equivalences

4.672. $Q \to (P \to R) \leftrightarrow (P \,\&\, Q) \to R \leftrightarrow P \to (Q \to R)$

are also immediate consequences of the associative law since they may be written in the form

$$\neg Q \vee (\neg P \vee R) \leftrightarrow (\neg P \vee \neg Q) \vee R \leftrightarrow \neg P \vee (\neg Q \vee R).$$

We come now to the distributive law

4.68. $\{P \vee (Q \,\&\, R)\} \to \{(P \vee Q) \,\&\, (P \vee R)\}$.

We have $Q \,\&\, R \to Q, \; Q \,\&\, R \to R$

and so

$$P \vee (Q \,\&\, R) \to P \vee Q, \qquad P \vee (Q \,\&\, R) \to P \vee R$$

whence

$$\{P \vee (Q \,\&\, R)\} \to \{(P \vee Q) \,\&\, (P \vee R)\}.$$

Before proving the converse of 4.68 we shall show that

4.681. $P \to (Q \to R) \vdash (A \vee P) \to \{(A \vee Q) \to (A \vee R)\}$.

By axiom 4.24

$$(Q \to R) \to \{(A \lor Q) \to (A \lor R)\}$$

and $\qquad\qquad A \to A \lor R,$

so that, by .441, $(A \lor Q) \to \{A \to (A \lor R)\}$

and therefore, by 4.672,

$$A \to \{(A \lor Q) \to (A \lor R)\}$$

whence we derive from $P \to (Q \to R)$

$$(A \lor P) \to \{(A \lor Q) \to (A \lor R)\}.$$

Since $Q \to (R \to (Q \& R))$ it follows that

$$(P \lor Q) \to \{(P \lor R) \to [P \lor (Q \& R)]\}$$

and using 4.672 again,

4.69. $\qquad \{(P \lor Q) \& (P \lor R)\} \to P \lor (Q \& R)$

which completes the proof of one of the distributive laws,

4.691. $\qquad P \lor (Q \& R) \leftrightarrow \{(P \lor Q) \& (P \lor R)\}.$

Replacing P, Q, R by their negations, and taking the negation of both sides of this equivalence (and replacing $\neg\neg P$ by P, etc.) we obtain the second distributive law

4.692. $\qquad P \& (Q \lor R) \leftrightarrow \{(P \& Q) \lor (P \& R)\}.$

4.7. Normal forms. Every sentence has an equivalent *conjunctive normal form*

$$P_1 \& P_2 \& \dots \& P_n$$

where each P_i is a disjunction of variables or negated variables. For example,

$$(p \& q) \lor \{(q \& r) \to p\}$$
$$\leftrightarrow (p \lor \neg q \lor \neg r) \& (p \lor q \lor \neg q \lor \neg r).$$

The normal form of a sentence is obtained by repeated application of the equivalences

$$\neg(P \,\&\, Q) \leftrightarrow \neg P \vee \neg Q, \qquad \neg(P \vee Q) \leftrightarrow \neg P \,\&\, \neg Q$$

which enable a negation sign to pass through any bracket until it stands in front of a variable, and the distributive law

$$\{P \vee (Q \,\&\, R)\} \leftrightarrow \{(P \vee Q) \,\&\, (P \vee R)\}$$

to remove a bracket round a conjunction.

Thus, for instance,

$$(p \,\&\, q) \vee \{(q \,\&\, r) \to p\} \leftrightarrow (p \,\&\, q) \vee \{\neg q \vee \neg r \vee p\}$$
$$\leftrightarrow (p \vee \neg q \vee \neg r \vee p) \,\&\, (q \vee \neg q \vee \neg r \vee p)$$
$$\leftrightarrow (p \vee \neg q \vee \neg r) \,\&\, (p \vee q \vee \neg q \vee \neg r).$$

Every sentence has also a second normal form, the disjunctive form,

$$Q_1 \vee Q_2 \vee \ldots \vee Q_n$$

where each Q_i is a conjunction of variables or negated variables. For if

$$\neg P \leftrightarrow P_1 \,\&\, P_2 \,\&\, \ldots \,\&\, P_n$$

then

$$P \leftrightarrow \neg P_1 \vee \neg P_2 \vee \ldots \vee \neg P_n$$

and $\neg P_i$ is a conjunction of variables or negated variables, since P_i is a disjunction of variables or negated variables.

4.71. If a sentence is universally true, then any equivalent sentence is universally true, for if P is true, and $P \to Q$ is true, then the truth table for implication shows that Q is also true (in fact $P \to Q$ is false only when P is true and Q is false). And if $P_1 \,\&\, P_2$ is universally true, then P_1 and P_2 are each universally true (for P_1 & P_2 is false unless *both* P_1 and P_2 are true). Thus if some sentence is universally true, and if P_1 & P_2 & ... & P_n is the normal form of the sentence then each of P_i is universally true. Suppose now that any variable which occurs negated in P_i does not also occur unnegated; then P_i cannot be universally true, for when we give the negated variables the value 0, and the unnegated variables the value 1, P_i takes the value 1. Thus if a sentence P is universally true, and if

$$P_1 \text{ \& } P_2 \text{ \& } ... \text{ \& } P_n$$

is its normal form, then each P_i contains *some* variable twice (at least), once without a negation prefixed, and also negated.

We have already remarked that $p \vee \neg p$ is provable for any variable p, and so

$$Q \vee p \vee \neg p \vee R$$

is provable, whatever sentences Q and R may be, so that P_i is provable. Hence, by 4.671, P_1 & P_2 & ... & P_n is provable, and so finally we see that P is provable.

Thus we have shown

Every universally true sentence is provable.

We express this fact by saying that axioms 4.21–4.24 are *complete* with respect to the truth tables. There is also another sense in which the axioms are complete: if we add to the axioms any sentence not provable from the axioms then the enlarged system is *inconsistent*, that is to say, every sentence becomes provable.

For if P is *not* provable from axioms 4.21–4.24, which we now call axiom system A, then the normal form of P,

$$P_1 \text{ \& } P_2 \text{ \& } ... \text{ \& } P_n$$

is not provable, and so, finally, some P_i is *not* provable, so that P_i does *not* contain the same variable negated and unnegated. Let us now take P as a fifth axiom and let us call this enlarged system of axioms A^+. Since P_i is derivable from P in A, it is provable in

A^+; replace each unnegated variable in P_i by p, and each negated variable by $\neg p$, so that P_i becomes $p \lor p \lor \ldots \lor p$, and therefore

$$p \lor p \lor \ldots \lor p$$

is provable in A^+, and so, by axiom 4.21, p itself is provable. But p is any sentence whatever, so that A^+ is inconsistent.

Axiom system A is consistent (not inconsistent) because only universally true sentences are provable in A, and so, for instance $p \And \neg p$ is *not provable*.

4.8. Axioms 4.21–4.24 are *independent*. To exhibit the independence of axiom 4.21 we set up a model for axioms 4.22, .23, .24 which does not satisfy 4.21. Let sentence variables take the values 0, 1, 2 and let $p \lor q$ have the value $p.q$ unless both p and q have the value 2, when we give $p \lor q$ the value 0. For the values 0, 1, 2 of p, $\neg p$ has the values 1, 0, 2 respectively. Then all the axioms except the first take the value 0, but $(2 \lor 2) \to 2$ has the value 2; moreover if P has the value 0, and $P \to Q$ has the value 0 then Q necessarily has the value 0. Of course any substitution in a sentence which takes only the value 0 leaves a sentence which only takes the value 0. Thus all the consequences of axioms 4.22–4.24 take only the value 0, but axiom 4.21 takes also the value 2, and so axiom 4.21 cannot be derived from axioms 4.22–4.24 and is independent of them.

To establish the independence of axiom 4.22 we give variables the same values 0, 1, 2 but now $p \lor q$ takes the value 2 when p and q have the value 2, and the values of $\neg p$ are 2, 1, 0 for the values 0, 1, 2 of p. Axioms 4.21, .23, .24 take only values 0, 1 but axiom 4.22 takes the value 2 (when p has the value 1 and q the value 2). If P does *not* take the value 2, and $P \to Q$ does *not* take the value 2 then Q does not take the value 2, and no sentence obtained from P by substitution takes the value 2, which proves that no sentence derived from 4.21, .23, .24 takes the value 2. Since axiom 4.22 can take the value 2, axiom 4.22 is *not* a consequence of the other axioms.

To exhibit the independence of the third axiom we consider a model with four values 0, 1, 2, 3 in which negation and disjunction are given by the tables

p	0	1	2	3
$\neg p$	1	0	0	2

$\vee$	1	2	3	q
1	1	2	3	
2	2	2	0	
3	3	3	3	
p				

$p \vee 0 = 0 \vee p = 0.$

In this model axioms 4.21, .22, .24 take only the value 0, and if $P = 0$ and $P \to Q = 0$ then $Q = 0$, but axiom 4.23 takes the value 3 when $p = 2$, $q = 3$.

The model for the proof of independence of the fourth axiom has the tables

p	0	1	2	3
$\neg p$	1	0	3	0

$\vee$	1	2	3	q
1	1	2	3	
2	2	2	0	
3	3	0	3	
p				

$p \vee 0 = 0 \vee p = 0$

In this model axioms 4.21, .22, .23 take only the value 0, axiom 4.24 takes the value 2 when $p = 3$, $q = 2$, $r = 2$, and from $P = 0$, $P \to Q = 0$ follows $Q = 0$.

4.9. We conclude this Chapter by proving the **Deduction theorem** for sentence logic.

4.91. *If Q is derivable from P, and if no substitution is made in variables in P in the derivation of Q, then $P \to Q$ is provable.*

Briefly: *if $P \vdash Q$ then $\vdash P \to Q$.*

Consider the derivation of Q from P. Each line in the derivation is either P or an axiom, or is inferred from a previous line by substitution, or is inferred from two previous lines by detachment. Prefix each line in this derivation by "$P \to$".

The first becomes $P \to P$ which is provable; an axiom A becomes $P \to A$ which is provable. An inference from B to C by substitution becomes an inference from $P \to B$ to $P \to C$ by substitution (since no substitution is made in variables in P) and an inference from $B, B \to C$ to C becomes an inference from $P \to B, P \to (B \to C)$ to $P \to C$, which is valid because

$$P \to (B \to C) \leftrightarrow B \to (P \to C)$$

and

$$P \to B, B \to (P \to C) \vdash P \to (P \to C)$$

and finally

$$P \to (P \to C) \leftrightarrow P \to C.$$

The last line of the proof is now $P \to Q$, so that $P \to Q$ is proved.

EXAMPLES IV

1. Prove that the following pairs of sentences are truth-table equivalent:
   .1 $p \to (q \& r), (p \to q) \& (p \to r)$
   .2 $p \to (q \lor r), (p \to q) \lor (p \to r)$
   .3 $p \to (q \to r), (p \to q) \to (p \to r)$
   .4 $(p \to q); p \to (p \to q)$
2. Verify the validity of the following sentences by the method of truth tables:
   .1 $(p \to r) \to \{(q \to r) \to [(p \lor q) \to r]\}$
   .2 $(p \to \neg p) \to \neg p$
   .3 $(\neg p \to p) \to p$
   .4 $(p \to q) \to \{(p \to \neg q) \to \neg p\}$
3. Derive the sentences 2.₁–2.4 above from axioms 4.21–4.24.
4. Express in conjunctive normal form the following sentences:
   .1 $\{(p \to q) \& (q \to p)\} \to \neg \{(p \to q) \lor (q \to p)\}$
   .2 $\{(p \to q) \lor r\} \to \{(q \to p) \& r\}$
   .3 $(\neg p \& \neg q) \to (r \to q)$
5. Prove the following 15 sentences:
   .1 $p \to (q \to p)$
   .11 $\{(p \to q) \to p\} \to p$
   .12 $(p \to q) \to \{(q \to r) \to (p \to r)\}$
   .2 $(p \& q) \to p$
   .21 $(p \& q) \to q$
   .22 $(p \to q) \to \{(p \to r) \to [p \to (q \& r)]\}$
   .3 $p \to (p \lor q)$
   .31 $q \to (p \lor q)$
   .32 $(p \to r) \to \{(q \to r) \to [(p \lor q) \to r]\}$

.4 $(p \leftrightarrow q) \rightarrow (p \rightarrow q)$
.41 $(p \leftrightarrow q) \rightarrow (q \rightarrow p)$
.42 $(p \rightarrow q) \rightarrow \{(q \rightarrow p) \rightarrow (p \leftrightarrow q)\}$
.5 $(p \rightarrow q) \rightarrow (\neg q \rightarrow \neg p)$
.51 $p \rightarrow \neg \neg p$
.52 $\neg \neg p \rightarrow p$

6. Prove the following rules of inference:
.1 $P, Q \vdash P \& Q$
.2 $P \vdash P \lor Q$
.3 $P \rightarrow Q, \neg Q \vdash \neg P$
.4 $(P \& Q) \rightarrow R, P \vdash Q \rightarrow R$

7. Show that every sentence has a truth table equivalent which employs the sole truth function $p|q$ with the truth table

p	q	$p|q$
0	0	1
0	1	1
1	0	1
1	1	0

8. Show that every sentence has a truth table equivalent which employs the sole truth function (p, q, r) with the truth table

p	q	r	(p, q, r)
0	0	0	1
0	0	1	1
0	1	0	0
0	1	1	0
1	0	0	0
1	0	1	0
1	1	0	1
1	1	1	0

CHAPTER FIVE

Lattices

A relation $a\ R\ b$ between the elements of a set K is said to be a *partial order* relation if

5.01. $a\ R\ a$ for every element a of K;

5.02. $a\ R\ b$ and $b\ R\ a$ imply $a = b$;

5.03. $a\ R\ b$ and $b\ R\ c$ imply $a\ R\ c$.

The set K is said to be partially ordered by the relation R. We do not require that one of the relations $a\ R\ b$, $b\ R\ a$ holds for every pair of elements. Thus for instance the relation of inclusion between classes is a partial order relation, for $A \subset A$ always holds, $A \subset B$ and $B \subset A$ imply that $A = B$, and if $A \subset B$, $B \subset C$ then $A \subset C$; but it is not true of any two classes that one of them is necessarily contained in the other. A relation R which satisfies 5.01, .02, .03 and is such that one of $a\ R\ b$, $b\ R\ a$ holds for *every* pair of elements a, b is called a relation of linear order, and K is said to be linearly ordered by R. For instance the natural numbers are linearly ordered by the relation "not greater than". A pair of elements a, b for which one of the relations $a\ R\ b$, $b\ R\ a$ holds are said to be *comparable*.

5.1. Since inclusion is the characteristic instance of a partial order relation we shall denote a partial ordering by the sign of inclusion $\subset$, and shall read $a \subset b$ as *a is in b* or *b contains a* (or *a is before b, b is after a*).

5.101. In a partially ordered set K, we have $a \subset b$ if and only if $k \subset a$ implies $k \subset b$ for every k in K. The necessity for the implication is obvious by 5.03; for the converse, we observe that a is one of the values of k, and so $a \subset a$ implies $a \subset b$.

Similarly we may show that $a \subset b$ if and only if $b \subset k$ implies $a \subset k$ for all k in K.

5.102. It follows from 5.101 that $a = b$ if and only if, for all k,

$$k \subset a \leftrightarrow k \subset b.$$

For $a \subset b$ if and only if $k \subset a$ implies $k \subset b$, for all k, and $b \subset a$ if and only if $k \subset b$ implies $k \subset a$.

5.11. An element a of a partially ordered set K which contains no element of K (other than itself) is said to be a *minimal element* of K. Thus for instance in the partial ordering of the human race by the relation "descended from" (reckoning a man amongst his own descendents) childless people are minimal elements. An element which is *contained in every element* is called a least or null element; if $0_1, 0_2$ are null elements then $0_1 \subset 0_2$ and $0_2 \subset 0_1$ so that $0_1 = 0_2$, and therefore the null element is unique. A partially ordered set may or may not have a null element; the empty class is the null element for classes under inclusion, but the positive fractions under the "not greater than" relation have no null element. When there is a null element this is necessarily the only minimal element. Elements which contain only themselves or the null element are called *atoms*.

5.12. An element which is not contained in any other element is called a *maximal* element; an element which contains every element is called the *unit* or greatest element (and the unit is unique); an element which is contained only in itself or in the unit element is called an *anti-atom*.

5.13. A linearly ordered subset of a partially ordered set is called a *chain*. Thus a (finite) chain is a set of elements $a_1, a_2, \ldots, a_n$ such that $a_1 \subset a_2 \subset a_3 \subset \ldots \subset a_n$.

5.2. Any *finite* chain has a first and a last element.

For if a is any element either there is an element a' such that $a' \subset a$, or $a \subset x$ for every element x; in the latter case a is the first element; in the former case either a' is the first element or there is an a'', $a'' \subset a'$. In every eventuality we reach a first element in a finite number of steps. In the same way we determine the last element of the chain.

5.21. Partially ordered sets with only a finite number of elements may be exhibited by means of an order diagram. For instance a set with elements a, b, c, d, e such that a, c, d are minimal, e is the unit element and a, b, e is a chain may be represented by the figure

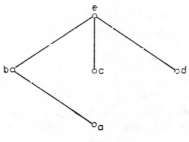

FIG. 1

In the figure inclusion is shown by an upward sloping link. Thus $a \subset b, b \subset e, c \subset e$ and $d \subset e$.

Another example is given by the figure

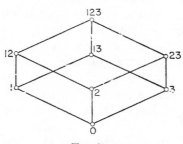

FIG. 2

where for instance 0, 1, 12, 123 and 0, 3, 13, 123 are chains. This diagram represents the class of all subclasses of the class with elements 1, 2, 3.

5.3. If H is a subset of a partially ordered set K and if there is an element u of K such that $h \subset u$ for every element h of H then u is called an *upper bound* of the set H. Note that u is comparable with every member of H. Similarly an element l of K such that $l \subset h$ for every element h of H is called a *lower bound* of H. If the set of upper bounds (a subset of K) has a least element this element is called the least upper bound of H, denoted by lub H; if the set of lower bounds has a greatest element, this element is called the greatest lower bound of H, and is denoted by glb H. Of course neither lub H nor glb H necessarily exists; a set in which lub H and glb H exist for every subset H is said to be *complete*. For example, in the partially ordered set exhibited in Fig. 2, the class of elements 0, 1, 2 has least upper bound 12, and the set 1, 2, 3 has least upper bound 123, but the subset of rationals x such that $x^2 < 2$ has no least upper bound in the set of rationals (for there is no rational whose square is 2).

If H is a pair of elements (a, b) and if the least upper and greatest lower bounds of H exist then of course

$$\text{lub}(a, b) = \text{lub}(b, a), \qquad \text{glb}(a, b) = \text{glb}(b, a).$$

Since glb H is a lower bound, we have glb $H \subset h$ for any element h of H; and so if $a \subset \text{glb } H$, then $a \subset h$ for all h in H, so that a is a lower bound of H. In fact

5.31. $\qquad x \subset \text{glb } H \leftrightarrow x \subset h$, for all h in H;

for if $x \subset h$, for all h, then x is a lower bound and so $x \subset \text{glb } H$, the *greatest* lower bound. Similarly

5.32. $\qquad \text{lub } H \subset y \leftrightarrow h \subset y$, for all h in H.

It follows in particular that

5.33. $\qquad \text{glb}(a, b) \subset a, \qquad \text{glb}(a, b) \subset b$

and

5.34. $\qquad a \subset \text{lub}(a, b), \qquad b \subset \text{lub}(a, b)$

and hence that

$$\text{glb}(a, a) = a \quad (\text{for } a \subset a \text{ and glb}(a, a) \subset a)$$
$$\text{lub}(a, a) = a \quad (\text{for } a \subset \text{lub}(a, a), a \subset a);$$

moreover

5.341.   if $a \subset b$ then $\text{glb}(a, b) = a$,

for a is a lower bound of a, b and $\text{glb}(a, b) \subset a$,
and

5.342. if $a \subset b$, $\text{lub}(a, b) = b$, for b is an upper bound of a, b
and $b \subset \text{lub}(a, b)$.

5.35. $\text{glb}(a, \text{glb}(b, c)) = \text{glb}(a, b, c)$

5.36. $\text{lub}(a, \text{lub}(b, c)) = \text{lub}(a, b, c)$.

By 5.31 we have

$$x \subset \text{glb}(a, b, c) \leftrightarrow x \subset a \ \& \ x \subset b \ \& \ x \subset c$$
$$\leftrightarrow x \subset a \ \& \ x \subset \text{glb}(b, c)$$
$$\leftrightarrow x \subset \text{glb}(a, \text{glb}(b, c))$$

whence 5.35 follows by 5.102.
5.36 follows similarly from 5.32.

5.37. $\text{glb}(a, \text{lub}(a, b)) = a$

5.38. $\text{lub}(a, \text{glb}(a, b)) = a$.

Since $a \subset \text{lub}(a, b)$, 5.37 follows from 5.341,
and since $\text{glb}(a, b) \subset a$, 5.38 follows from 5.342.
5.4. A partially ordered set in which each pair of elements (a, b)
has a least upper bound and greatest lower bound is called a
lattice. In a lattice we denote the least upper bound of a, b by
$a \cup b$, and the greatest lower bound by $a \cap b$.

We proceed now to show that in any lattice the commutative,
associative and contraction laws hold:

5.41. $a \cup b = b \cup a$; $a \cap b = b \cap a$.

5.42. $(a \cup b) \cup c = a \cup (b \cup c)$; $(a \cap b) \cap c = a \cap (b \cap c)$.

5.43. $a \cap (a \cup b) = a$; $a \cup (a \cap b) = a$.

The relations 5.41 are implicit in the definition of the least upper and greatest lower bound (§ 5.3); 5.42 follows immediately from 5.35, 5.36; and 5.43 follows from 5.37, 5.38.

5.5. Conversely a set closed under two operations $\cup$, $\cap$ which satisfy 5.41, .42, .43 is a lattice. We define

5.51. $a \subset b \leftrightarrow a \cap b = a;$

it follows from 5.43 (and 5.41) that

$$(b \cap a) \cup b = b, \qquad a \cap (a \cup b) = a$$

and so $a \subset b$ is also equivalent to $a \cup b = b$.

We have to show that $a \cup b$, $a \cap b$ defined by 5.41, .42, .43, are the least upper and greatest lower bounds of a, b with respect to the inclusion relation defined in 5.51.

First we must prove that $a \subset b$ is a relation of partial order, that is, we must show

$a \subset a; \qquad a \subset b \,\&\, b \subset c \to a \subset c; \qquad a \subset b \,\&\, b \subset a \to a = b.$

Using 5.51 these relations become

$$a \cap a = a; \qquad a \cap b = a \,\&\, b \cap c = b \to a \cap c = a;$$
$$a \cap b = a \,\&\, b \cap a = b \to a = b.$$

The third of these follows at once from 5.41.

For the first we note that by 5.43

$$a = a \cap (a \cup (a \cap b)) = a \cap a, \quad \text{and for the second,}$$

if $a \cap b = a \quad \text{and} \quad b \cap c = b \quad \text{then}$

$$a \cap c = (a \cap b) \cap c = a \cap (b \cap c) = a \cap b = a.$$

We come now to the proof that $a \cup b$, $a \cap b$ are the least upper and greatest lower bounds of a, b.

We have to show

5.52. $a \cap b \subset a, \qquad a \cap b \subset b;$

5.53. $a \subset a \cup b, \qquad b \subset a \cup b;$

5.54. $k \subset a \,\&\, k \subset b \to k \subset a \cap b;$

5.55. $a \subset k \,\&\, b \subset k \to a \cup b \subset k.$

For 5.52 we observe that $a \cap b \subset a \leftrightarrow (a \cap b) \cup a = a$, and
$(a \cap b) \cup a = a \cup (a \cap b) = a$ by 5.41, .43. Similarly $a \cap b \subset b$.
Again, for 5.53 we require $a \cap (a \cup b) = a$, $b \cap (a \cup b) = b$
which follow from 5.43 (and .41).
If $k \subset a$ and $k \subset b$ then $k \cap a = k$ and so $k \cap (a \cap b)$
$= (k \cap a) \cap b = k \cap b = k$, which proves 5.54.
Similarly if $a \subset k$ and $b \subset k$ then

$$(a \cup b) \cup k = a \cup (b \cup k) = a \cup k = k, \quad \text{which proves 5.55,}$$

and completes the proof of 5.5.

We conclude this section by noting that in a lattice with zero
and unit 0, 1, we have

$$0 \subset a, \qquad a \subset 1 \quad \text{for all } a,$$

and so

$$a \cup 0 = a, \qquad a \cap 0 = 0; \qquad a \cup 1 = 1, \qquad a \cap 1 = a.$$

5.6.   A lattice in which the two distributive laws

5.61. $a \cap (b \cup c) = (a \cap b) \cup (a \cap c)$
5.62. $a \cup (b \cap c) = (a \cup b) \cap (a \cup c)$

hold is called a *distributive lattice*.

For instance, the lattice of all subclasses (of a given class) is
distributive, as we saw in Chapter I. But the lattice given by this
diagram

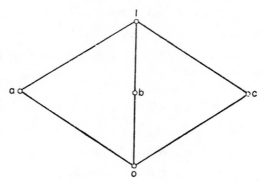

is not distributive, since $a \cup (b \cap c) = a$ but

$$(a \cup b) \cap (a \cup c) = 1 \cap 1 = 1.$$

In a distributive lattice the two equations

$$a \cap b = a \cap c, \qquad a \cup b = a \cup c$$

imply that $\qquad\qquad\qquad b = c.$

For $\quad b = b \cup (a \cap b) = b \cup (a \cap c) = (b \cup a) \cap (b \cup c)$
$\qquad = (a \cup c) \cap (b \cup c) = (a \cap b) \cup c = (a \cap c) \cup c = c.$

A distributive lattice (with unit) may also be characterized by axiom 5.61, the axioms

5.621. $\qquad\qquad (b \cup c) \cap a = (b \cap a) \cup (c \cap a)$

5.63. $\qquad\qquad\qquad a \cap a = a$

and by postulating the existence of an element 1 such that

5.64. $\qquad\qquad\qquad a \cup 1 = 1 \cup a = 1,$

5.65. $\qquad\qquad\qquad a \cap 1 = 1 \cap a = a,$

in place of the commutative, associative and contraction laws. We start by proving the dual of .63, namely $a \cup a = a$; in fact

$$a = a \cap 1 = a \cap (a \cup 1) = (a \cap a) \cup (a \cap 1) = a \cup a.$$

For the contraction laws we have

$$(a \cap b) \cup a = (a \cap b) \cup (a \cap 1) = a \cap (b \cup 1) = a \cap 1 = a$$

and

$$a \cup (a \cap b) = (a \cap 1) \cup (a \cap b) = a \cap (1 \cup b) = a \cap 1 = a$$

and similarly

$$a \cup (b \cap a) = (b \cap a) \cup a = a.$$

For the second contraction law we have

$$a \cap (a \cup b) = (a \cap a) \cup (a \cap b) = a \cup (a \cap b) = a$$

and

$$a \cap (b \cup a) = (a \cap b) \cup (a \cap a) = (a \cap b) \cup a = a,$$
$$(a \cup b) \cap a = (a \cap a) \cup (b \cap a) = a \cup (b \cap a) = a,$$
$$(b \cup a) \cap a = (b \cap a) \cup (a \cap a) = (b \cap a) \cup a = a.$$

By means of the contraction laws we may now proceed to prove the commutative law

$$a \cup b = b \cup a.$$

We have

$$
\begin{aligned}
a \cup b &= \{a \cap (b \cup a)\} \cup \{b \cap (b \cup a)\} \\
&= (a \cup b) \cap (b \cup a), \qquad \text{by .621,} \\
&= \{(a \cup b) \cap b\} \cup \{(a \cup b) \cap a\} \\
&= b \cup a, \qquad \text{by the contraction laws.}
\end{aligned}
$$

We prove next some extended forms of the contraction laws:

$$a \cap \{(a \cup b) \cup c\} = a$$
$$b \cap \{(a \cup b) \cup c\} = b$$
$$c \cap \{(a \cup b) \cup c\} = c.$$

For the first of these

$$a \cap \{(a \cup b) \cup c\} = \{a \cap (a \cup b)\} \cup (a \cap c) = a \cup (a \cap c) = a,$$

and similarly

$$b \cap \{(a \cup b) \cup c\} = \{b \cap (a \cup b)\} \cup (b \cap c) = b \cup (b \cap c) = b,$$
$$
\begin{aligned}
c \cap \{(a \cup b) \cup c\} &= \{c \cap (a \cup b)\} \cup (c \cap c) \\
&= \{c \cap (a \cup b)\} \cup c = c.
\end{aligned}
$$

The associative law for union now follows, for

$$
\begin{aligned}
a \cup (b \cup c) &= [a \cap \{(a \cup b) \cup c\}] \cup ([b \cap \{(a \cup b) \cup c\}] \cup \\
&\qquad\qquad\qquad\qquad\qquad \cup [c \cap \{(a \cup b) \cup c\}]) \\
&= [a \cap \{(a \cup b) \cup c\}] \cup [(b \cup c) \cap \{(a \cup b) \cup c\}], \\
&\qquad\qquad\qquad\qquad\qquad\qquad\qquad\qquad \text{by .621} \\
&= \{a \cup (b \cup c)\} \cap \{(a \cup b) \cup c\}, \qquad \text{again by .621;}
\end{aligned}
$$

5.7.

next we observe that the following extended contraction laws may be proved in the same way

$$\{a \cup (b \cup c)\} \cap a = a$$
$$\{a \cup (b \cup c)\} \cap b = b$$
$$\{a \cup (b \cup c)\} \cap c = c$$

and from these we have

$$a \cup b = [\{a \cup (b \cup c)\} \cap a]$$
$$\cup [\{a \cup (b \cup c)\} \cap b]$$
$$= \{a \cup (b \cup c)\} \cap (a \cup b)$$

and so

$$(a \cup b) \cup c = [\{a \cup (b \cup c)\} \cap (a \cup b)]$$
$$\cup [\{a \cup (b \cup c)\} \cap c]$$

5.71. $$= \{a \cup (b \cup c)\} \cap \{(a \cup b) \cup c\}$$

and so from 5.7, 5.71 we reach

$$a \cup (b \cup c) = (a \cup b) \cup c.$$

The second distributive law may now be proved. We have

$$(a \cup b) \cap (a \cup c) = \{a \cap (a \cup c)\} \cup \{b \cap (a \cup c)\}, \qquad \text{by } .621,$$
$$= a \cup \{b \cap (a \cup c)\}$$
$$= a \cup \{(b \cap a) \cup (b \cap c)\}, \qquad \text{by } .61,$$
$$= \{a \cup (b \cap a)\} \cup (b \cap c),$$
$$= a \cup (b \cap c)$$

which proves 5.62 (the commutative law for union showing that

$$(b \cap c) \cup a = (b \cup a) \cap (c \cup a)).$$

For the commutative law for intersection we have now

$$a \cap b = \{a \cup (b \cap a)\} \cap \{b \cup (b \cap a)\}$$
$$= (a \cap b) \cup (b \cap a)$$
$$= \{(a \cap b) \cup b\} \cap \{(a \cap b) \cup a\}$$
$$= b \cap a.$$

Finally the proof of the associative law for intersection may now be written down simply by interchanging union and intersection in the proof of the associative law for union. This completes the proof that axioms 5.61, .621, .63, .64, .65 determine a distributive lattice with unity.

5.8. If to each element x in a lattice with zero and unit, corresponds an element x' such that

$$x \cap x' = 0, \qquad x \cup x' = 1$$

the lattice is said to be *complemented*.

If the lattice is distributive the complement of x is unique for if x has two complements x' and X we have

$$x \cap x' = x \cap X = 0$$

$$x \cup x' = x \cup X = 1$$

and since the lattice is distributive it follows that $X = x'$, and the complement is unique.

A complemented distributive lattice is a Boolean algebra, and conversely, for each is closed under two operations $\cap$, $\cup$ which are commutative, associative and distributive, each over the other; in each there are unique elements 0, 1 and in each an element x has a unique complement x' such that

$$x \cup x' = 1, \qquad x \cap x' = 0,$$

and

$$x \cup 0 = x, \qquad x \cap 1 = x.$$

5.81. A (non-empty) subclass $I_\cup$ of a lattice K is called a *union-ideal* of the lattice if for any a, b of $I_\cup$, $a \cup b$ belongs to $I_\cup$, and for any k of K, $a \cap k$ belongs to $I_\cup$. That is

$$a \, \varepsilon \, I_\cup \, \& \, b \, \varepsilon \, I_\cup \rightarrow a \cup b \, \varepsilon \, I_\cup$$

and

$$a \, \varepsilon \, I_\cup \, \& \, k \, \varepsilon \, K \rightarrow a \cap k \, \varepsilon \, I_\cup.$$

In other words the elements of an ideal are closed under union, and the intersection of a member of the ideal with any member of the lattice is itself a member of the ideal.

A (non-empty) subclass $\mathscr{I}_\cap$ of a lattice K is called an intersection-ideal if for any a, b of $\mathscr{I}_\cap$, $a \cap b$ belongs to $\mathscr{I}_\cap$, and for any k of K, $a \cup k$ belongs to $\mathscr{I}_\cap$.
In symbols

$$a \, \varepsilon \, \mathscr{I}_\cap \, \& \, b \, \varepsilon \, \mathscr{I}_\cap \to a \cap b \, \varepsilon \, \mathscr{I}_\cap$$

$$a \, \varepsilon \, \mathscr{I}_\cap \, \& \, k \, \varepsilon \, K \to a \cup k \, \varepsilon \, \mathscr{I}_\cap.$$

5.82. A non-empty subclass I of a lattice K which is closed under union is a union-ideal if and only if I contains any element contained in a member of I. That is to say I is a union-ideal if I is closed under union and

$$a \, \varepsilon \, I \, \& \, b \subset a \to b \, \varepsilon \, I.$$

For if $b \subset a$ then $b = a \cap b$ and $a \cap b$ is contained in a union-ideal when a is contained in the ideal.
Conversely, if

$$a \, \varepsilon \, I \, \& \, b \subset a \to b \, \varepsilon \, I$$

and if k is any element of K, then $a \cap k \subset a$ and so $a \cap k \, \varepsilon \, I$, proving that I is a union-ideal.

5.83. A non-empty subclass I of a lattice K is a union-ideal if and only if

$$a \, \varepsilon \, I \, \& \, b \, \varepsilon \, I \leftrightarrow a \cup b \, \varepsilon \, I.$$

One half of this equivalence is already contained in the definition of a union-ideal, and if I is a union-ideal such that

$$a \cup b \, \varepsilon \, I \quad \text{then} \quad a = (a \cup b) \cap a$$

so that

$$a \, \varepsilon \, I,$$

and similarly

$$b \, \varepsilon \, I.$$

It remains to show that a subclass closed under union, and which contains $a \cup b$ only if it contains both a and b, is a union-ideal; let k be any member of K, then we must show that the intersection of k with any member a of the subclass I is itself a member of the subclass. In fact

$$a = (a \cap k) \cup a$$

so that, since a itself belongs to I, therefore $a \cap k$ belongs to I, as was to be shown.

5.84. The dual of 5.82 affirms that a non-empty subclass $\mathscr{I}$ closed under intersection is an intersection-ideal if and only if $\mathscr{I}$ contains any element which contains a member of $\mathscr{I}$, that is, if:

5.841. $a \, \varepsilon \, \mathscr{I} \, \& \, a \subset b \to b \, \varepsilon \, \mathscr{I}$.

For if $a \subset b$ then $b = a \cup b$ and so if $\mathscr{I}$ is an intersection-ideal and $a \, \varepsilon \, \mathscr{I}$ then $b \, \varepsilon \, \mathscr{I}$. Conversely if $\mathscr{I}$ is closed under intersection and satisfies .841, and if $a \, \varepsilon \, \mathscr{I}$ and if k is any member of K then

$$a \subset a \cup k$$

and so $a \cup k \, \varepsilon \, \mathscr{I}$ and $\mathscr{I}$ is an intersection-ideal.

5.85. The dual of 5.83 says that a non-empty subclass $\mathscr{I}$ of a lattice K is an intersection-ideal if and only if

5.851. $a \, \varepsilon \, \mathscr{I} \, \& \, b \, \varepsilon \, \mathscr{I} \leftrightarrow a \cap b \, \varepsilon \, \mathscr{I}$.

If $\mathscr{I}$ is an intersection-ideal then $a \, \varepsilon \, \mathscr{I} \, \& \, b \, \varepsilon \, \mathscr{I} \to a \cap b \, \varepsilon \, \mathscr{I}$; and if $a \cap b \, \varepsilon \, \mathscr{I}$ then $a = (a \cap b) \cup a$ so that $a \, \varepsilon \, \mathscr{I}$, and similarly $b \, \varepsilon \, \mathscr{I}$. Conversely if $\mathscr{I}$ is a subclass satisfying .851 then $\mathscr{I}$ is closed under intersection, and if k is any element of K, and a any element of $\mathscr{I}$, then since,

$$(a \cup k) \cap a = a$$

it follows that $a \cup k \, \varepsilon \, \mathscr{I}$ and so $\mathscr{I}$ is an intersection-ideal.

5.86. An ideal of a lattice is itself a lattice, for the ideal is partially ordered by the inclusion relation of the lattice and both $a \cup b$, $a \cap b$ belong to the ideal when each of a, b belongs to the ideal (whether the ideal is an intersection or union-ideal).

5.87. The set of union ideals of a lattice L is a lattice in which the lattice-intersection of two ideals is just their common part and the lattice-union of two ideals A, B is the class of elements of L contained in the union of an element of A and an element of B.

We remark first that the common part of two union-ideals A, B is a union-ideal, for if $a \varepsilon A \cap B$ and $b \varepsilon A \cap B$ then since a, b belong to A, B, therefore $a \cup b$ belongs to both A and B and so to their common part; and if c is any element of L then $a \cap c$ belongs to A and to B, and so to their common part. $A \cap B$ is necessarily the largest ideal which is contained in A and B.

Next we show that the set K of all k in L such that $k \subset a \cup b$ for an element a of A, and an element b of B, is a union-ideal in L. For if $k_1 \varepsilon K$ and $k_2 \varepsilon K$ then there are elements a_1, a_2 of A, and elements b_1, b_2 of B such that

$$k_1 \subset a_1 \cup b_1, \qquad k_2 \subset a_2 \cup b_2$$

and therefore

$$k_1 \cup k_2 \subset (a_1 \cup b_1) \cup (a_2 \cup b_2) = (a_1 \cup a_2) \cup (b_1 \cup b_2);$$

since $a_1 \cup a_2 \varepsilon A$ and $b_1 \cup b_2 \varepsilon B$ therefore $k_1 \cup k_2 \varepsilon K$. Moreover, if $k \varepsilon K$ and $d \subset k$ then $d \subset a \cup b$ for some $a \varepsilon A$, and some $b \varepsilon B$, so that $d \varepsilon K$ which proves that K is a union-ideal. Since $a \subset a \cup b$, and $b \subset a \cup b$, for any $a \varepsilon A$, $b \varepsilon B$ therefore $a \varepsilon K$ and $b \varepsilon K$ so that $A \subset K$ and $B \subset K$. If we denote the *lattice-union* of two ideals A, B (as just defined) by $A \cup B$ (to avoid confusion with set union) we have to show that the lattice axioms are satisfied. Since we have taken lattice intersection to be just set intersection necessarily the commutative and associative laws for intersection are satisfied. So too since $a \cup b = b \cup a$, we have $A \cup B = B \cup A$. We proceed to prove the associative law

$$(A \cup B) \cup C = A \cup (B \cup C).$$

$(A \cup B) \cup C$ is the class of all elements x of L such that $x \subset k \cup c$ where $c \varepsilon C$ and $k \subset a \cup b$ for some $a \varepsilon A, b \varepsilon B$. Then $x \subset a \cup (b \cup c)$ and so $x \subset a \cup l$ where $l = b \cup c$ and so $l \varepsilon B \cup C$ which proves that $x \, \varepsilon \, A \cup (B \cup C)$. Similarly any member of $A \cup (B \cup C)$ is a member of $(A \cup B) \cup C$ which completes the proof of the associative law.

There remain to be proved the two contraction laws

$$A \cap (A \cup B) = A, \qquad A \cup (A \cap B) = A.$$

For the first of these we observe that $A \subset A \cup B$ and so $A \subset A \cap (A \cup B)$; since necessarily $A \cap (A \cup B) \subset A$, the first equation is proved. For the second we remark first that $A \subset A \cup (A \cap B)$; and if $x \, \varepsilon \, A \cup (A \cap B)$ then $x \subset a \cup u$ for some $a \, \varepsilon \, A, u \, \varepsilon \, A \cap B$. Then since a, u both belong to the union-ideal A, therefore $a \cup u$ belongs to A and hence x belongs to A, which completes the proof of the second contraction law.

5.88. If L is a distributive lattice the lattice union $K = A \cup B$ of two ideals A, B is equal to the set U of all $a \cup b$ for an $a \, \varepsilon \, A$ and a $b \, \varepsilon \, B$. Since $a \cup b \subset a \cup b$ therefore $U \subset K$; and if $x \, \varepsilon \, K$ then $x \subset a \cup b$ for some $a \, \varepsilon \, A, b \, \varepsilon \, B$, and so

$$x = x \cap (a \cup b) = (x \cap a) \cup (x \cap b) = a_1 \cup b_1$$

where $a_1 = x \cap a, b_1 = x \cap b$ so that $a_1 \, \varepsilon \, A, b_1 \, \varepsilon \, B$
and therefore $x \, \varepsilon \, U$, proving that $U = K$.

5.89. If L is a distributive lattice then the lattice of the ideals of L is also distributive.

Let A, B, C be ideals in L, then we have to prove

5.891. $$A \cap (B \cup C) = (A \cap B) \cup (A \cap C)$$

5.892. $$A \cup (B \cap C) = (A \cup B) \cap (A \cup C)$$

For .891 it suffices to prove the inclusion

$$A \cap (B \cup C) \subset (A \cap B) \cup (A \cap C)$$

since the opposite inclusion

$$(A \cap B) \cup (A \cap C) \subset A \cap (B \cup C)$$

holds in any lattice (see example V, 1.92).

Let $a \,\varepsilon\, A \cap (B \cup C)$ then $a \,\varepsilon\, A$ and $a = b \cup c$
for some $b \,\varepsilon\, B$, $c \,\varepsilon\, C$; hence

$$a = a \cap a = a \cap (b \cup c) = (a \cap b) \cup (a \cap c)$$

which belongs to $(A \cap B) \cup (A \cap C)$ as was to be shown.
The second distributive law follows from the first (see example
V, 2).

5.9. *Any* **finite** *Boolean algebra $\mathscr{B}$ is isomorphic to the algebra of
all subsets of the set of atoms of the algebra.* We recall that an
atom of a partially ordered set is an element $a \neq 0$ such that only
0 and a itself is contained in a. The only point in the proof at
which we use the finiteness of $\mathscr{B}$ is to show that $\mathscr{B}$ has atoms. (It
is a consequence of our succeeding results that an infinite Boolean
algebra may fail to have atoms.) Consider any element $b \neq 0$;
then there is an atom contained in b. For either b is an atom or
there is a non-zero element b_1, contained in b, and $b \neq b_1$; either
b_1 is an atom or there is a non-zero element b_2 contained in b_1,
and $b_2 \neq b_1$. Either b_2 is an atom or it contains an element b_3
and so on. Since $\mathscr{B}$ is finite the sequence $b, b_1, b_2, \ldots$ must
terminate at some atom b_k contained in b.

Atoms divide the elements of $\mathscr{B}$ into two separate classes, for if
a is an atom and b is any element of $\mathscr{B}$, then a is either contained
in b, or in b' (and not in both, since from $a \subset b$ and $a \subset b'$
follows $b \subset a'$ and so $a \subset a'$, that is $a = a \cap a' = 0$). For
$a \cap b \subset a$, and a is an atom, and so either $a \cap b = 0$ (that is
$a \subset b'$) or $a \cap b = a$, and so $a \subset b$.

Let $A(b)$ denote the set of all the atoms of $\mathscr{B}$ which are contained
in b.
We shall show that the sets $A(b)$ comprise exactly the subsets
of the set of atoms (i.e. that every subset is one of the sets $A(b)$).
We start by proving that, for any elements b, c

$$A(b \cup c) = A(b) \cup A(c).$$

Let a be an atom of the set $A(b \cup c)$ so that $a \subset b \cup c$; then $a \subset b$ or $a \subset c$ so that a is contained in (at least) one of the sets $A(b)$, $A(c)$ and so is contained in their union. Conversely if $a \, \varepsilon \, A(b) \cup A(c)$ then either $a \, \varepsilon \, A(b)$ or $a \, \varepsilon \, A(c)$, that is either $a \subset b$ or $a \subset c$ and therefore $a \subset b \cup c$, that is, $a \, \varepsilon \, A(b \cup c)$. Next we observe that, if a is an atom, $A(a)$ has the single member a. Consider now any subset $(a_1, a_2, \dots , a_k)$ of the set of atoms, and let $b = a_1 \cup a_2 \cup \dots \cup a_k$. Then

$$A(b) = A(a_1) \cup A(a_2) \cup \dots \cup A(a_k) = (a_1, a_2, a_3, \dots , a_k).$$

Thus every subset of the set of atoms is expressible in the form $A(b)$. This correspondence between b and $A(b)$ is one-to-one, for if $b \neq c$, then either b is not contained in c, or c is not contained in b (since $c \subset b \, \& \, b \subset c \to b = c$); suppose that b is not contained in c, then $b \cap c' \neq 0$, so that there is an atom a contained in $b \cap c'$ and so contained in both b, c' (and not therefore in c) from which it follows that $a \, \varepsilon \, A(b)$ and $a \notin A(c)$ and so $A(b) \neq A(c)$.

Finally we show that the correspondence between b and $A(b)$ is an isomorphism. We have already seen that $A(b \cup c) = A(b) \cup A(c)$ so that union is preserved under the correspondence. It remains to show that $A(b')$ is the complement (in the set of atoms) of $A(b)$; in fact $a \, \varepsilon \, A(b') \leftrightarrow a$ is not contained in $b \leftrightarrow a \notin A(b) \leftrightarrow a$ is contained in the complement of $A(b)$, as was to be proved.

Since a finite class of n elements has precisely 2^n subclasses (for each subclass either does, or does not, contain each element) it follows that the number of elements of a finite Boolean algebra is a power of 2.

5.91. It is not true that an infinite Boolean algebra is isomorphic to the set of all subsets of a set. Instead we have **Stone's theorem** that *an infinite Boolean algebra is isomorphic to some family of subsets of a set*, as we now proceed to prove. We shall see that the place of atoms, in the proof, is taken by certain ideals.

Since we shall be concerned only with one kind of ideal in the proof, the union-ideal, we shall speak simply of *ideals*, dropping the qualification. We consider the set $\mathscr{S}$ of all ideals of a Boolean algebra $\mathscr{B}$ which do not contain 1. The elements of $\mathscr{S}$, being subclasses, are partially ordered by the relation of inclusion; with respect to this partial order $\mathscr{S}$ may have maximal elements, and we shall call a maximal element of $\mathscr{S}$ a *maximal ideal*. To secure the existence of a maximal ideal we are obliged to introduce an axiom which (as far as is known) is independent of all other set axioms which we have used (tacitly) up to this point. The axiom in question, in a form which it was given by M. Zorn, is this: **If every chain in a partially ordered set has an upper bound then the set has a maximal element.**

This axiom is known to be equivalent to the *axiom of choice* which may be expressed by saying that the class of all subclasses, omitting the null class, of a set S, may be mapped into S in such a way that the map of each subclass B belongs to B, or equivalently by saying that for any family F of sets without common elements there is a set which contains exactly one member of each set of the family.

Let $\mathscr{C}$ be any chain in $\mathscr{S}$, and form the class $\mathscr{A}$ of all elements which belong to some ideal in the chain $\mathscr{C}$. That is to say, an element of $\mathscr{B}$ belongs to $\mathscr{A}$ if and only if there is an ideal of $\mathscr{B}$ in the chain $\mathscr{C}$ which contains this element. If I is any ideal in $\mathscr{C}$ then every member of I is a member of $\mathscr{A}$, so that I is contained in $\mathscr{A}$, and therefore $\mathscr{A}$ is an upper bound of the chain of ideals $\mathscr{C}$. Moreover $\mathscr{A}$ itself is an ideal, for if a and b belong to $\mathscr{A}$, then there are ideals I, J of $\mathscr{C}$ such that $a \, \varepsilon \, I, b \, \varepsilon \, J$. Since $\mathscr{C}$ is a chain, one of I, J is contained in the other, so that a, b are elements of the same ideal, which therefore contains also $a \cup b$; since $a \cup b$ belongs to an ideal of $\mathscr{C}$, therefore $a \cup b$ belongs to $\mathscr{A}$. Furthermore, if k is any member of $\mathscr{B}$, since I is an ideal, $a \cap k$ belongs to I and so to $\mathscr{A}$. Thus $\mathscr{A}$ is an ideal. Finally we remark that 1 does not belong to any ideal in $\mathscr{S}$ and so does not belong to $\mathscr{A}$, which proves that $\mathscr{A}$ is an ideal of the set $\mathscr{S}$. Thus every chain in $\mathscr{S}$ has an upper bound in $\mathscr{S}$, and so, by Zorn's axiom, $\mathscr{S}$ has a maximal element.

By confining attention to a subclass of $\mathscr{S}$, consisting of ideals which contain some given ideal $\mathscr{I}_0$, we observe that the above proof determines a maximal ideal which includes $\mathscr{I}_0$. In particular therefore we can find a maximal ideal which contains any chosen element $b \neq 1$, for we can find a maximal ideal which contains the ideal (b) of all elements x of B such that $x \subset b$.

Maximal ideals separate the members of $\mathscr{B}$ into two distinct classes; given a maximal ideal P and any element b of $\mathscr{B}$ then either $b \varepsilon P$ or $b' \varepsilon P$. For suppose that $b \notin P$ and consider the class X of all elements of the form $x \cup p$, where p is any element of P and x is any element of $\mathscr{B}$ such that $x \subset b$. We shall show that X is an ideal. If x_1, x_2 are both contained in b, and if p_1, p_2 are any two elements of P then

$$(x_1 \cup p_1) \cup (x_2 \cup p_2) = (x_1 \cup x_2) \cup (p_1 \cup p_2)$$

so that $(x_1 \cup p_1) \cup (x_2 \cup p_2)$ is a member of X; and if b_1 is any member of $\mathscr{B}$ then

$$(x_1 \cup p_1) \cap b_1 = (x_1 \cap b_1) \cup (p_1 \cap b_1)$$

and since $x_1 \cap b_1 \subset x_1 \subset b$, and $p_1 \cap b_1$ belongs to P, therefore $(x_1 \cup p_1) \cap b_1$ is a member of X, which completes the proof that X is an ideal. Now any member p of P belongs to X, for $p = (p \cap x) \cup p$ and $p \cap x \subset b$ if $x \subset b$, and yet X is not equal to P for $b = b \cup 0$ and $b \subset b$ and $0 \varepsilon P$ so that $b \varepsilon X$, whereas $b \notin P$. Since P is a maximal element of $\mathscr{S}$, and $P \subset X$, therefore X is an ideal which is not a member of $\mathscr{S}$, and so an ideal which contains 1. But if an ideal contains 1 it contains $1 \cap a = a$ for any element a of $\mathscr{B}$, so that in fact $\mathscr{B}$ itself is the only ideal which contains 1. Thus $X = \mathscr{B}$, and so every element a is expressible in the form $x \cup p$ for some $x \subset b$ and some $p \varepsilon P$. In particular there is a p such that $b \cup p = 1$; but the equation $b \cup p = 1$ entails $p = b'$ which proves that $b' \varepsilon P$. Of course it is not possible for both b and b' to belong to P for then we should have $1 = b \cup b'$ as a member of the ideal P.

We proceed now to consider properties of subsets of the set of all maximal ideals. Let $M(x)$ be the set of all maximal ideals P such that $x' \varepsilon P$. We shall show that the set $\mathscr{E}$ of sets $M(x)$, for all x in $\mathscr{B}$, is closed under set-intersection and complementation.

The intersection of two sets $M(x)$, $M(y)$ is again a member of $\mathscr{E}$; for in fact $M(x) \cap M(y) = M(x \cap y)$. For if $P \varepsilon M(x \cap y)$ then P contains $(x \cap y)' = x' \cup y'$; but P contains $x' \cup y'$ only if x' and y' each belong to P, and therefore P belongs to both $M(x)$ and $M(y)$ and so to their intersection $M(x) \cap M(y)$. Conversely if P belongs to both $M(x)$ and $M(y)$ then x' and y' belong to P, so that $x' \cup y' = (x \cap y)'$ belongs to P, and therefore $P \varepsilon M(x \cap y)$.

The complement of $M(x)$ in $\mathscr{E}$ is $M(x')$, for $P \varepsilon M(x') \leftrightarrow x \varepsilon P \leftrightarrow P \notin M(x)$.

Thus $\mathscr{E}$ is closed under intersection and complementation and so forms a Boolean algebra.

The sets $M(x)$ and $M(y)$ are equal if and only if $x = y$; for if $x \neq y$ then either x is not contained in y or y is not contained in x. Without loss of generality we may suppose that x is not contained in y, and therefore $x \cap y' \neq 0$, whence $x' \cup y \neq 1$, and so there is a maximal ideal P which contains $x' \cup y$, and therefore contains both x' and y. Since $x' \varepsilon P$, $P \varepsilon M(x)$ and since $y \varepsilon P$, $P \varepsilon M(y')$; thus P belongs to $M(x)$ but not to $M(y)$, which proves that $M(x) \neq M(y)$.

It follows now that $\mathscr{B}$ is isomorphic to $\mathscr{E}$; for there is a one-one correspondence between x and $M(x)$, under which the mate of $x \cap y$ is $M(x \cap y) = M(x) \cap M(y)$, and the mate of x' is $M(x')$, the complement in $\mathscr{E}$ of $M(x)$.

We have proved Stone's theorem that every Boolean algebra is isomorphic to the algebra of a family of subsets of the set of maximal ideals.

$\mathscr{E}$ will not in general contain all the subsets of the set of maximal ideals, for if $\mathscr{B}$ is infinite there are, by Stone's theorem, an infinity of subsets of the set of maximal ideals, and so the set of maximal ideals is infinite. We may readily show that the set of all subsets of an infinite set is not denumerable. For suppose that $S_1, S_2, S_3, \ldots$

is an enumeration of *all* subsets of the set of natural numbers
1, 2, 3, ... , and form the subset S as follows: S contains 1 only if
S_1 does not; S contains 2 only if S_2 does not, and so on. Then S
is a subclass different from each of S_1, S_2, ... and so S is a subclass
not contained in the enumeration. To complete the proof that $\mathscr{E}$
does not in general contain all the subsets of the set of maximal
ideals, we shall show that there are denumerably infinite Boolean
algebras.

Consider the family $\mathscr{F}$ of all subsets of natural numbers which are
either finite (including the empty set) or, if infinite, contain all
natural numbers from some number onwards (and perhaps
other numbers as well).

Thus for instance $\mathscr{F}$ contains the sets

$$\{1, 2, 3\}, \qquad \{1, 5, 7, 11\}, \qquad \{21, 22, 23, ...\}$$
$$\{1, 5, 7, 11, 21, 22, 23, ...\}.$$

Each infinite set is the complement (in the set of natural numbers)
of a finite set. All finite sets of natural numbers may be enumerated,
the set with elements

$$a_1 < a_2 < a_3 < ... < a_k$$

being assigned the position

$$2^{a_1 - 1} + 2^{a_2 - 2} + ... + 2^{a_k - 1}.$$

(Thus, for instance, since $13 = 2^3 + 2^2 + 2^0$, the thirteenth set
in the enumeration is 1, 3, 4.) If we denote the sequence of finite
sets by $f_1, f_2, ...$ then the infinite sets are $f_1', f_2', ...$ and the whole
family $\mathscr{F}$ may be enumerated as $f_1, f_1', f_2, f_2', ...$.

To show that $\mathscr{F}$ is a Boolean algebra of sets we have only to
show that $\mathscr{F}$ is closed under intersection and complementation.
Since the infinite sets are the complements of the finite sets in $\mathscr{F}$,
and conversely, $\mathscr{F}$ is closed under complementation. Moreover
the intersection of two finite sets is a finite set, the intersection of
two infinite sets in $\mathscr{F}$ is an infinite set in $\mathscr{F}$, and finally the
intersection of a finite and of an infinite set is a finite set, therefore
$\mathscr{F}$ is closed under complementation.

5.92. Newman algebra. A generalization of a Boolean algebra, obtained by dropping the commutative and associative axioms, is called a Newman algebra. In a Newman algebra we have a set N closed under two operations, which we denote by $a + b$ and ab, such that

5.921. $a(b + c) = ab + ac,$ $(a + b)c = ac + bc.$

5.922. There is an element 1 such that $a1 = a$, for all a.

5.923. There is an element 0 such that $a + 0 = a = 0 + a$, for all a.

5.924. To each a corresponds at least one complement a' such that

$$aa' = 0, \qquad a + a' = 1.$$

We start by proving that

5.93. $aa = a,$

5.94. $(a')' = a.$

For .93 we have

$$aa = aa + 0 = aa + aa' = a(a + a') = a1 = a.$$

For .94 we have

$$
\begin{aligned}
(a')' &= 0 + (a')'(a')' \\
&= a'(a')' + (a')'(a')' \\
&= (a' + (a')')(a')' \\
&= 1(a')' = (a + a')(a')' = a(a')' + 0 \\
&= 0 + a(a')' = aa' + a(a')' \\
&= a(a' + (a')') = a1 = a.
\end{aligned}
$$

It follows that $a'a = 0$ and $a' + a = 1$, and also that

$$1a = a,$$

for $1a = (a + a')a = aa + a'a = a + 0 = a.$

From this we readily prove that complements are unique; for if a' and $a*$ are both complements of a then

$$
\begin{aligned}
a* &= a*1 = a*(a' + a) = a*a' + a*a \\
&= a*a' + 0 = a*a' + aa' = (a* + a)a' \\
&= 1a' = a'.
\end{aligned}
$$

Since $1 + 0 = 1$, and complements are unique, $0 = 1'$ and $0' = 1$. Next we observe that $a0 = 0 = 0a$ for all a, for

$$0 = aa' = a(a' + 0) = aa' + a0 = 0 + a0 = a0,$$

and

$$0 = bb' = (0 + b)b' = 0b' + bb' = 0b' + 0 = 0b',$$

whence taking $b = a'$, for any a, so that $b' = a$,

we have $$0a = 0.$$

If $0 = 1$ it follows that

$$0 = 0 + 0 = 0 + a0 = 0 + a1 = a1 = a$$

for any a, and so we suppose that $0 \neq 1$.

5.95. If we define $2 = 1 + 1$ then

$$2 + 2 = 2.1 + 2.1 = 2(1 + 1) = 2.2 = 2.$$

We call the left multiples of 2, $a2$, *even* elements.
An element a is even if and only if $a + a = a$,
for if $a = b2$ then $a + a = b(2 + 2) = b2$,
and if $a + a = a$ then

$$a = a + a = a1 + a1 = a(1 + 1) = a2$$

so that a is even. *Thus addition of even elements is idempotent.*
Any left or right multiple of an even number is even, for if a is even, so that $a = a + a$ then

$$ab = (a + a)b = ab + ab, \qquad \text{so that } ab \text{ is even,}$$

and

$$ba = b(a + a) = ba + ba \qquad \text{so that } ba \text{ is even.}$$

We remark next that

5.951. $(a + b)2 = a2 + b2$, $(ab)2 = (a2)(b2)$, $(a2)2 = a2$,

so that the class of even elements is closed under addition and multiplication.

The first of these is just an instance of the distributive law. For the second we have

$$(a2)(b2) = (a + a)(b + b) = (a + a)b + (a + a)b$$
$$= (ab + ab) + (ab + ab)$$
$$= (ab)2 + (ab)2$$
$$= (ab)2,$$

and for the third

$$(a2)2 = a2 + a2 = a(2 + 2) = a2.$$

Further, if a is even,

$$a + 1 = (a + 1)1 = (a + 1)(a + a') = (aa + a) + (aa' + a')$$
$$= (a + a) + (0 + a') = a + a' = 1$$

and similarly $1 + a = 1$.

Hence $a2 + 2 = a2 + 1.2 = (a + 1)2 = 1.2 = 2$, and similarly $2 + a2 = 2$.

We observe now that the even elements satisfy the axioms 5.61, .621, .63, .64 and .65 with addition taking the part of union and multiplication taking the part of intersection so that the even elements form a distributive lattice with unit 2; in this lattice $a'2$ is the complement of $a2$ since $a2 + a'2 = (a + a')2 = 2$, and $(a2)(a'2) = (aa')2 = 0$, and the complement of 2 is 0. Thus the even elements form a complemented distributive lattice, that is, a Boolean algebra.

5.96. We conclude this account of Newman algebra by proving that addition is both commutative and associative.

We observe first that

5.97. If $a + b = 0$ then $a = b$ and $b + a = 0$.

For
$$a = a(b' + b) = ab' + ab = (ab' + bb') + ab$$
$$= (a + b)b' + ab = 0b' + ab = ab,$$

and
$$b = (a' + a)b = (a'a + a'b) + ab = ab$$

so that $a = b$ and
$$b + a = a + b = 0.$$

5.98. Let $c = (a + b)'$ and $d = (b + a)'$, for any a, b, then
$$a + b = (a + b)1 = (a + b)(d + d')$$
$$= (a + b)d + c'd';$$

but $0 = (b + a)d = bd + ad$ so that $(a + b)d = ad + bd = 0$, by .97, and therefore
$$a + b = c'd'.$$

Similarly
$$b + a = 1(b + a) = (c + c')(b + a)$$
$$= c(b + a) + c'd';$$

but $0 = c(a + b) = ca + cb$, so that $c(b + a) = cb + ca = 0$ and so
$$b + a = c'd'$$

which proves that
$$a + b = b + a.$$

5.99. We prove the associative law in three steps, starting with the proofs of two special cases of the law. In each proof the method is to establish an equation $l = r$ by showing that $al = ar$, and $a'l = a'r$ from which it follows that

$$l = (a' + a)l = a'l + al = a'r + ar = (a' + a)r = r.$$

The first case we consider is

5.991. $1 + (1 + a) = (1 + 1) + a;$

writing l and r for the left and right hand sides of this equation, we have
$$al = a + a(1 + a) = a + (a + a) = (a + a) + a, \qquad \text{by .98,}$$
$$= ar,$$

and
$$a'l = a' + a'(1 + a) = a' + a' = a'r$$

which proves .991. Next we consider

5.992. $\qquad 1 + (a + b) = (1 + a) + b;$

again denote the left and right hand sides of this equation by l, r then

$$al = a + a(a + b) = a + (a + ab) = a\{1 + (1 + b)\}$$
$$= a\{(1 + 1) + b\}, \qquad \text{by .991,}$$
$$= (a + a) + ab = ar,$$

and

$$a'l = a' + a'(a + b) = a' + a'b = a'r,$$

which proves .992.

Finally, denoting $a + (b + c)$ by l, and $(a + b) + c$ by r we have

$$al = a\{a + (b + c)\} = a + a(b + c) = a\{1 + (b + c)\}$$
$$= a\{(1 + b) + c\}, \qquad \text{by .992,}$$
$$= ar,$$

and

$$a'l = a'(b + c) = a'\{(a + b) + c\} = a'r$$

so that $l = r$, that is

$$a + (b + c) = (a + b) + c.$$

EXAMPLES V

1. Prove that in any lattice
.1 $a \cap a = a \cup a = a$
.2 $a \cap b = a \leftrightarrow a \cup b = b$
.3 $a \cap b = a \cup b \rightarrow a = b$
.4 $a \cap b \cap c = a \cup b \cup c \rightarrow a = b = c$
.5 $c \subset a \,\&\, c \subset b \rightarrow c \subset a \cap b$
.6 $a \subset c \,\&\, b \subset c \rightarrow a \cup b \subset c$
.7 $a \subset b \rightarrow a \cap c \subset b \cap c$
.8 $a \subset b \rightarrow a \cup c \subset b \cup c$
.9 $a \subset b \,\&\, c \subset d \rightarrow a \cap c \subset b \cap d$
.91 $a \subset b \,\&\, c \subset d \rightarrow a \cup c \subset b \cup d$
.92 $(a \cap b) \cup (a \cap c) \subset a \cap (b \cup c)$
.93 $a \cup [(a \cup b) \cap (a \cup c)] = (a \cup b) \cap (a \cup c)$
2. Prove that in a lattice in which one distributive law holds the second distributive law also holds.

3. A mapping $a \to \bar{a}$ of a lattice L on to a lattice $\bar{L}$ is called a *lattice homomorphism* if

$$a \subset b \to \bar{a} \subset \bar{b}$$

and a union (intersection) homomorphism if

$$\overline{a \cup b} = \bar{a} \cup \bar{b} \qquad (\overline{a \cap b} = \bar{a} \cap \bar{b}).$$

Prove that a union (intersection) homomorphism is a lattice homomorphism.

4. For a fixed element k prove that the mappings

$$a \to a \cap k$$
$$a \to a \cup k$$

are lattice homomorphisms.

5. A lattice is said to be complete if every set of elements has a least upper and greatest lower bound. Prove that a homomorphic mapping of a complete lattice on to itself leaves at least one element invariant.

6. Show that in any lattice

$$(a \cap b) \cup (b \cap c) \cup (c \cap a) \subset (a \cup b) \cap (b \cup c) \cap (c \cup a).$$

7. Prove that a lattice is distributive if and only if

7.1 $(a \cap b) \cup (b \cap c) \cup (c \cap a) = (a \cup b) \cap (b \cup c) \cap (c \cup a).$

8. Prove that in a distributive lattice

$$a \cap x = a \cap y \,\&\, a \cup x = a \cup y \to x = y.$$

9. If k is a fixed element of a lattice L show that the set of elements contained in k forms a union-ideal, called the principal ideal generated by k, and denoted by (k). Prove that in a finite lattice every ideal is a principal ideal.

10. If J is the set of positive integers and if for all positive integers m, n we define $m \cap n$ to be the highest common factor, and $m \cup n$ the least common multiple of m, n prove that J is a lattice with respect to these two operations.

11. If (a) and (b) are principal ideals in the lattice of ideals of a lattice L prove that

$$(a) \cap (b) = (a \cap b), \qquad (a) \cup (b) = (a \cup b).$$

12. Prove that the mapping of a lattice L upon the set of principal ideals of the lattice of ideals of L is an isomorphism.

13. A set of elements R is called a *ring* with respect to two operations $a + b$, ab if, for all elements $a, b, a + b$ and ab belong to R, and

$$a + b = b + a, \qquad a + (b + c) = (a + b) + c$$
$$(ab)c = a(bc)$$
$$a(b + c) = ab + ac, \qquad (b + c)a = ba + ca$$

and there is an element 0, and corresponding to each a an element $\bar{a}$ such that

$$a + 0 = a, \qquad a + \bar{a} = 0.$$

If $ab = ba$ the ring is said to be commutative, and if there is an element 1 such that $a \cdot 1 = 1 \cdot a = a$ the ring is said to be a ring with unit. A *Boolean* ring is a ring with unit in which $aa = a$ for all elements a. Prove that in a Boolean ring $a + a = 0$, and that every Boolean ring is commutative.

14. If a, b are elements of a Boolean ring and if

$$a \cap b = ab, \qquad a \cup b = a + b + ab$$
$$a' = a + 1$$

prove that the elements of the ring form a Boolean algebra with operations $\cap, \cup, '$.

15. Prove that a Boolean algebra is a Boolean ring with respect to symmetric difference and intersection.

16. Prove that the lattice

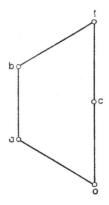

is not distributive, but that the two lattices

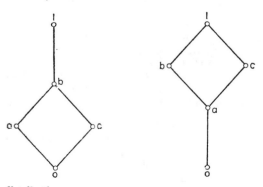

are both distributive.

Solutions

Examples I

1. $A' \cup B = 1$ is equivalent to $(A' \cup B)' = 1'$, i.e. to $A \cap B' = 0$, which by 1.85 is equivalent to $A \subseteq B$.

2. $(A \cup B \cup C)' = \{(A \cup B) \cup C\}' = (A \cup B)' \cap C' = A' \cap B' \cap C'$
$(A \cap B \cap C)' = \{(A \cap B) \cap C\}' = A' \cup B' \cup C'$.
The general results are
$$(A_1 \cup A_2 \cup ... \cup A_n)' = A_1' \cap A_2' \cap ... \cap A_n'$$
$$(A_1 \cap A_2 \cap ... \cap A_n)' = A_1' \cup A_2' \cup ... \cup A_n',$$
(proof by induction).

3. By 1.82, $X \subseteq A \cap A' = 0$; but $0 \subseteq X$ so that $X = 0$, and $1 = A \cup A' \subseteq X$ so that $1 \subseteq X$; but $X \subseteq 1$, proving $X = 1$. If $A \subseteq B$ then $B = A \cup B$, and if $C \subseteq D$, $D = C \cup D$ and so $B \cup D = A \cup B \cup C \cup D = (A \cup C) \cup (B \cup D)$ so that $A \cup C \subseteq B \cup D$.

4. In fact $A \cap B \subseteq A$, $A \cap B \subseteq B$ and if $A \cap B = 0$ then
$$A = A \cap (B \cup B') = (A \cap B) \cup (A \cap B') = A \cap B'$$
so that $A \subseteq B'$; thus if A is not contained in B', $A \cap B \neq 0$.

5. $A + B = (A \cap B') \cup (A' \cap B)$, $A' + B' = (A' \cap B) \cup (A \cap B')$.

6. Add K to both sides forming $A + K + K = B + K + K$, whence since $K + K = 0$, the result follows.
Similarly from $A + B = 0$ follows $A + B + B = B$, i.e. $A = B$.

7. $(A + B) + (C + D) = \{(A + B) + C\} + D$, by the associative law,
$$= \{A + (B + C)\} + D$$
$$= \{A + (C + B)\} + D$$
$$= \{(A + C) + B\} + D$$
$$= (A + C) + (B + D).$$

8. $(A + B)' = \{(A \cap B') \cup (A' \cap B)\}' = \{(A' \cup B) \cap (A \cup B')\}$
$$= \{(A' \cup B) \cap A\} \cup \{(A' \cup B) \cap B'\}$$
$$= (A \cap B) \cup (A' \cap B') = A' + B = A + B'.$$
$(A - K) \cup (B - K) = (A \cap K') \cup (B \cap K') = (A \cup B) \cap K'$
$$= (A \cup B) - K.$$
$\{(A + K) \cup (B + K)\}' = (A + K') \cap (B + K')$
$$= (A \cap B) + K' \cap (A + B) + K' \cap 1$$
$$= (A \cap B) + K' \cap (A + B)', \quad \text{since}$$
$$A + B + 1 = (A + B)',$$
and so, taking complements
$$(A + K) \cup (B + K) = (A \cap B) + K \cup (A + B).$$

122

9. $A + (A \cup B) = (A \cap A' \cap B') \cup (A' \cap (A \cup B))$
$\qquad = A' \cap B,$
$\quad B + (A \cap B) = (B \cap (A' \cup B')) \cup (B' \cap A \cap B)$
$\qquad = B \cap A'.$
$\quad B - (A \cap B) = B \cap (A' \cup B') = B \cap A'.$
From $A + (A \cup B) = B + (A \cap B)$
follows $A + B + (A \cap B) = A + A + (A \cup B) = A \cup B.$

10. From 9.

11. $(A + B) \times C = (A + B) + C' = (A + C') + B = (A \times C) + B$
$\qquad\qquad\qquad\qquad = (B + C') + A = (B \times C) + A.$
$\quad (A \times C) + (B \times C) = (A + C') + (B + C') = A + B.$

12.1. From 9, $B + (A \cup B) = B' \cap A = A - B$ whence $(A - B) + B =$
$\quad B + B + (A \cup B) = A \cup B.$

.2. $(A - B) \cap B = A \cap B' \cap B = 0.$

.3. $A \cap (A - B) = A \cap A \cap B' = A \cap B' = A - B.$

.4. $A - B = A \cap B' \subset A.$

.5. $A - A = A \cap A' = 0.$

.6. $A - (B - C) = A \cap (B \cap C')' = A \cap (B' \cup C) = (A \cap B') \cup (A \cap C).$

.7. $A - (A - B) = A \cap (A \cap B')' = A \cap (A' \cup B) = A \cap B.$

.8. $(A - B) - C = A \cap B' \cap C',$
$\quad (A - C) - (B - C) = (A \cap C') \cap (B \cap C')'$
$\qquad\qquad\qquad\qquad = (A \cap C') \cap (B' \cup C)$
$\qquad\qquad\qquad\qquad = A \cap B' \cap C'.$

.9. $A - (B \cap C) = A \cap (B \cap C)' = A \cap (B' \cup C')$
$\qquad\qquad\qquad\qquad = (A \cap B') \cup (A \cap C').$

.91. $A - (B \cup C) = A \cap B' \cap C',$
$\qquad\qquad (A - B) \cap (A - C) = A \cap B' \cap A \cap C' = A \cap B' \cap C'.$

.92. $(A \cup B) - B = (A \cup B) \cap B' = A \cap B'.$
$\quad A - (A \cap B) = A \cap (A' \cup B') = A \cap B'.$

.93. $(A \cup B) \cup (B - A) = A \cup B \cup (B \cap A') = (A \cup B) \cap 1 = A \cup B.$

13.1. If $A = 0$ and $B = 0$, $A \cup B = 0 \cup 0 = 0$; and if $A \cup B = 0$ then
$\quad A \subset A \cup B = 0$ and so since $0 \subset A$, $A = 0$. Similarly $B = 0.$

.2. If $A \cap B' = A$ then $A' \cup (A \cap B') = A' \cup A = 1$, so that $A' \cup B' = 1,$
$\quad$ whence $B = B \cap (A' \cup B') = B - A.$

.3. If $A \cup B = A \cap B'$ then $B = (A \cup B) \cap B = A \cap B' \cap B = 0.$

.4. If $A \cap B = A \cap B'$ then
$\quad A = A \cap 1 = A \cap (B + B')$
$\qquad = (A \cap B) + (A \cap B') = (A \cap B) + (A \cap B) = 0.$

.5. $A \subset A \cup B$ and so, if $A \cup B \subset C$, then $A \subset C$. If $A \subset C$ and $B \subset C,$
$\quad$ then $(A \cup B) \cap C' = (A \cap C') \cup (B \cap C') = 0.$

.6. Since $A \cap B \subset A$ therefore $C \subset A \cap B$ implies $C \subset A$; if $C \subset A$ and
$\quad C \subset B$ then $C \cap (A \cap B)' = (C \cap A') \cup (C \cap B') = 0.$

.7. $A \subset B \cup C$ is equivalent to $A \cap (B \cup C)' = 0$ which is equivalent to
$\quad A \cap B' \cap C' = 0.$

.8. If $A - B = B - A$ then $A = A \cup (A \cap B') = A \cup (B \cap A') = A \cup B;$
$\quad$ and $B = B \cup (B \cap A') = B \cup (A \cap B') = B \cup A.$

.9. $A = A \cup (A \cap B) = A \cup (A \cup B) = A \cup B$, if $A \cup B = A \cap B$, and $B = B \cup (A \cap B) = B \cup (A \cup B) = A \cup B$.

.91. If $A \subset B \subset C$ then $A \cup B = B$, $B \cap C = B$; if $A \cup B = B \cap C$ then $B = B \cap (A \cup B) = B \cap B \cap C = B \cap C$, so that $B \subset C$, and $A \cup B = B$ so that $A \subset B$.

.92. If $A \cap B' = 0$ and $C \cap D' = 0$ then $(A \cap B') \cup (C \cap D') = 0$; if $(A \cap B') \cup (C \cap D') = 0$ then $A \cap B' = 0$, $C \cap D' = 0$.

.93. If $A + B = 0$ then $A = A + 0 = A + A + B = B$.
     If $A = B$ and $C = D$ then $A + B = 0$, $C + D = 0$ and so $(A + B) \cup (C + D) = 0$; conversely if $(A + B) \cup (C + D) = 0$ then $A + B = 0$, $C + D = 0$, and so $A = B$, $C = D$.

.94. If $A \cap X' = B \cap X'$ then $(A + B) \cap X' = (A \cap X') + (B \cap X') = (A \cap X') + (A \cap X') = 0$; if $(A + B) \cap X' = 0$ then $(A \cap X') + (B \cap X') = 0$ and so $A \cap X' = B \cap X'$.

14.1. $(A \cup B) \cap (B \cup C) \cap (C \cup A) = \{B \cup (A \cap C)\} \cap (C \cup A)$
$$= \{B \cap (C \cup A)\} \cup \{(A \cap C) \cap (A \cup C)\}$$
$$= (A \cap B) \cup (B \cap C) \cup (C \cap A).$$

.2. From .1, $(A \cap B) \cup (A \cap C) \cup (A \cap D) \cup (B \cap C) \cup (B \cap D) \cup (C \cap D)$
$$= \{(A \cup B) \cap (B \cup C) \cap (C \cup A)\} \cup \{(A \cup B \cup C) \cap D\}$$
$$= (A \cup B \cup C) \cap [\{(A \cup B) \cap (B \cup C) \cap (C \cup A)\} \cup D]$$
$$= (A \cup B \cup C) \cap (B \cup C \cup D) \cap (C \cup D \cup A) \cap (D \cup A \cup B).$$

.3. $A - (B \cup C) = A \cap B' \cap C' = (A - B) - C$.

.4. $(A - B) \cap C = A \cap B' \cap C = (A \cap C) - B$.

.5. $(A \cup B) - C = (A \cup B) \cap C' = (A - C) \cup (B - C)$.

.6. $A - (B - A) = A \cap (B \cap A')' = A \cap (A \cup B') = A$.

.7. $(A - C) \cap (B - C) = (A \cap C') \cap (B \cap C') = A \cap B \cap C'$.

.8. $(A - B) \cap (C - D) = A \cap B' \cap C \cap D' = (A \cap C) \cap (B \cup D)'$.

.9. $A - [B - (C - D)] = A \cap [B \cap (C' \cup D)]'$
$$= A \cap [B' \cup (C \cap D')]$$
$$= (A - B) \cup [A \cap C \cap D'].$$

.91. $(A \cap B') \cup (B \cap C') \cup (C \cap A') \cup (A \cap B \cap C)$
$$= \{(A \cap B') \cup (A \cap B \cap C)\} \cup (B \cap C') \cup (C \cap A')$$
$$= [A \cap (A \cup B) \cap (A \cup C) \cap (B' \cup C)] \cup (B \cap C') \cup (C \cap A')$$
$$= [A \cap (B \cap C')'] \cup (B \cap C') \cup (C \cap A')$$
$$= A \cup (B \cap C') \cup (C \cap A')$$
$$= (A \cup C) \cup (B \cap C') = A \cup B \cup C.$$

15. The set consists of the classes 0, A, B, $A \cap B$, $A - B$, $B - A$.

16. $A \cup C \subset (A \cup C) \cup B = (A \cup B) \cup (B \cup C)$.
$$[(A \cap B') \cup (B \cap C')] \cap (A \cap C') = [(A \cap B') \cup (A \cap B \cap C')] \cap C'$$
$$= (A \cap B' \cap C') \cup (A \cap B \cap C')$$
$$= (A \cap C') \cap (B \cup B') = A \cap C'.$$

Hence
$$A - C \subset (A - B) \cup (B - C)$$
$$C - A \subset (B - A) \cup (C - B)$$
and so
$$A + C = (A - C) \cup (C - A) \subset (A - B) \cup (B - A) \cup (B - C) \cup (C - B)$$
$$= (A + B) \cup (B + C)$$

17. By 16,

$$A - C \subseteq (A - B) \cup (B - C),$$
$$A - D \subseteq (A - C) \cup (C - D)$$

whence

$$(A - C) \cup (C - D) \subseteq (A - B) \cup (B - C) \cup (C - D)$$

and so

$$A - D \subseteq (A - B) \cup (B - C) \cup (C - D).$$

Examples II

1.1. By the distributive law.

.2. By repeated application of the distributive law.

.3. From .2 by complementation.

.4. $\{(A \cup B) \cap (A' \cup C)\} \cup \{(A \cup D) \cap (A' \cup E)\}$
$= (A \cup B \cup A \cup D) \cap (A \cup B \cup A' \cup E) \cap (A' \cup C \cup A \cup D)$
$\quad \cap (A' \cup C \cup A' \cup E)$
$= (A \cup B \cup D) \cap (A' \cup C \cup E).$

.5. $(A \cap B') + (A' \cap B) = \{A \cap B' \cap (A \cup B')\} \cup \{(A' \cup B) \cap (A' \cap B)\}$
$\qquad\qquad = (A \cap B') \cup (A' \cap B) = A + B$

.6. The complement of $(A + B) \cup (B + C)$ is

$$(A' + B) \cap (B' + C) = A' \cap B' + (A' + B) \cap C$$

and the complement of $(A + C) \cup (B + C)$ is

$$(A' + C) \cap (B' + C) = A' \cap B' + C \cap (A' + B' + 1)$$
$$= A' \cap B' + C \cap (A' + B),$$

since $A' + B' + 1 = (A' + B')' = A' + B.$

.7. Take $C = 0$ in .6.

.8. $A + (A \cup B) = \{A' \cap (A \cup B)\} \cup \{A \cap (A' \cap B')\}$
$\qquad\qquad = B \cap A' = B - A.$

.9. We have $B - A = B \cap A' = B \cap (A' \cup B') = B - (A \cap B)$ and so
$A + \{B - (A \cap B)\} = A + (B - A) = A + A + (A \cup B),$ by .8,
$\qquad\qquad\qquad = A \cup B.$

.91. By .6.

.92. $(A \cup B) \cap (A' \cup C) = (A \cap A') \cup (A' \cap B) \cup (A \cap C) \cup (B \cap C)$
$= [(B \cap C) \cap (A \cup A')] \cup (A' \cap B) \cup (A \cap C)$
$= [(A \cap C) \cup \{(A \cap C) \cap B\}] \cup [(A' \cap B) \cup \{(A' \cap B) \cap C\}]$
$= (A \cap C) \cup (A' \cap B).$

2.

A	B	B'	$A \cap B'$	$(A \cap B') \cap B'$	$(A \cap B')' \cap B$	$(A - B) + B$	$A \cup B$
0	0	1	0	0	0	0	0
0	1	0	0	0	1	1	1
1	0	1	1	1	0	1	1
1	1	0	0	0	1	1	1

The agreement of the last two columns proves $(A - B) + B = A \cup B.$

3. $(A \cup B)' \cap A \cap B = A' \cap B' \cap A \cap B$.

  $(A \cup B \cup C)' \cap (A \cup B) = A' \cap B' \cap C' \cap (A \cup B)$
  $$= (A' \cap B' \cap C' \cap A) \cup (A' \cap B' \cap C' \cap B)$$

  which are the two normal forms.

  $(A' \cap B)' \cap (C \cup B')' \cap (C \cup A') = (A \cup B') \cap B \cap C' \cap (A' \cup C)$
  $= (A \cap A' \cap B \cap C') \cup (A \cap B \cap C \cap C') \cup (A' \cap B \cap B' \cap C')$
  $\quad \cup (B \cap B' \cap C \cap C')$.

4. To A corresponds $A + K$ and if $A + K = B + K$ then $A = B$, so that the correspondence is one-to-one. We have $(A + K)' = A' + K$, so that complements are preserved.

5. $(A, B)' = (A', B') \to (B')'$
  $(A, B) \cap (C, D) = (A \cap C, B \cup D) \to (B \cup D)' = B' \cap D'$
  $(A, B) \cup (C, D) = (A \cup C, B \cap D) \to (B \cap D)' = B' \cup D'$.

6. See § 2.96.

7. See § 2.96.

8. If A and B map upon the same element then $A \cup B$, $A \cap B$ also map on this element.

9. We have $A \subset B$ if and only if $B = A \cup B$;
  under the mapping $A \to A^* \cup \bar{A}$ we have

  $$B \to B^* \cup \bar{B}, A \cup B \to (A \cup B)^* \cup \overline{(A \cup B)} = A^* \cup B^* \cup \bar{A} \cup \bar{B}$$
  $$= A^* \cup \bar{A} \cup B^* \cup \bar{B}$$

  and so, if $A \subset B$, and if $M(A)$, $M(B)$ are the maps $A^* \cup \bar{A}$, $B^* \cup \bar{B}$, then $M(A) \subset M(B)$. For the mapping $A \to A^* \cap \bar{A}$ we write $A \subset B$ in the form $A = A \cap B$.

10. Let $K(a)$ be the class of prime factors of the number a, and $K'(a)$ the class of prime factors of N which are not factors of a. Then $K(a) = K(b)$ if and only if $a = b$; $K(N)$ plays the part of the universal class, and $K(1)$ is the null class. Since $a \cup b$ is the product of the prime factors of a and b, each factor counted once only, and $a \cap b$ is the product of the primes common to a and b, therefore

  $$K(a \cup b) = K(a) \cup K(b), \qquad K(a \cap b) = K(a) \cap K(b);$$

  since $a' = N/a$, therefore $K(a') = K'(a)$.

  From $K(a \cup b) = K(a) \cup K(b) = K(b) \cup K(a) = K(b \cup a)$ follows $a \cup b = b \cup a$, and similarly we prove $a \cap b = b \cap a$.

  From

  $$K(a \cup (b \cap c)) = K(a) \cup K(b \cap c) = K(a) \cup \{K(b) \cap K(c)\}$$
  $$= \{K(a) \cup K(b)\} \cap \{K(a) \cup K(c)\}$$
  $$\text{(by the distributive law for classes)}$$
  $$= K(a \cup b) \cap K(a \cup c)$$
  $$= K((a \cup b) \cap (a \cup c))$$

  whence $a \cup (b \cap c) = (a \cup b) \cap (a \cup c)$
  and similarly $a \cap (b \cup c) = (a \cap b) \cup (a \cap c)$.
  Next we note that

  $$K(a \cup 1) = K(a) \cup K(1) = K(a) \cup 0 = K(a)$$

so that $\qquad a \cup 1 = a,$
and
$$K(a \cap N) = K(a) \cap K(N) = K(a) \cap 1 = K(a)$$
so that $\qquad a \cap N = a.$
Finally, we note that
$$K(a \cup a') = K(a) \cup K(a') = K(a) \cup K'(a) = K(N)$$
$$K(a \cap a') = K(a) \cap K(a') = K(a) \cap K'(a) = K(1)$$
so that $a \cup a' = N$, $a \cap a' = 1$ completing the proof.

11. $B \cap A = (C \cup A) \cap A = (C \cap A) \cup A = 0 \cup A = A,$
and
$B \cap C = (C \cup A) \cap C = C \cup (A \cap C) = C \cup 0 = C;$
Since
$$C \cap (C \cup A) = C \cup (C \cap A) = C$$
and $\qquad D \cap (D \cup A) = D$
therefore from $\qquad C \cup A = D \cup A \qquad$ we have
$$C = C \cap (C \cup A) = C \cap (D \cup A) = C \cap D$$
$$D = D \cap (D \cup A) = D \cap (C \cup A) = D \cap C$$
so that $\quad C = D.$

12. We have $X \cup A = (A' \cap B) \cup A = (A' \cap B) \cup (A \cap B) = (A' \cup A) \cap B = B,$
and $\quad X \cap A = B \cap A' \cap A = 0.$

The solution is unique by the previous example.

13. We have
$$\begin{aligned}
X \cup (A \cap B) &= \{X \cap (A \cup B)\} \cup (A \cap B) \\
&= (X \cap A) \cup (X \cap B) \cup (A \cap B) \\
&= \{(X \cap A) \cup (B \cap A)\} \cup \{(X \cap B) \cup (A \cap B)\} \\
&= \{(X \cup B) \cap A\} \cup \{(X \cup A) \cap B\} \\
&= A \cup B
\end{aligned}$$
and
$$X \cap (A \cap B) = 0, \qquad (A \cap B) \cap (A \cup B) = A \cap B,$$
so that by the previous example $X = (A \cup B) \cap (A \cap B)' = A + B$.
Conversely, if $X = A + B$, then
$$\begin{aligned}
X \cap (A \cup B) &= (A + B) \cap \{(A + B) + A \cap B\} \\
&= A + B + (A + B) \cap (A \cap B) \\
&= A + B = X.
\end{aligned}$$
$$A \cap (B \cup X) = A \cap \{B \cup (A + B)\} = A \cap \{B \cup A\} = A,$$
by example 1.7,
$$B \cap (A \cup X) = B \cap \{A \cup (A + B)\} = B \cap \{A \cup B\} = B,$$
and
$$\begin{aligned}
X \cap A \cap B &= (A + B) \cap A \cap B \\
&= (A \cup B) \cap (A \cap B)' \cap (A \cap B) = 0.
\end{aligned}$$

Examples III

8. By 6, 3 $\qquad (A' \cup B)' \cup (A' \cup B)' = A$

and so, by 2, taking B' for B,
$$(A' \cup B')' \cup (A' \cup B')' = A$$

and 8 now follows from 7.

9. $A \cup A' = [(A' \cup A''')' \cup (A' \cup A'')'] \cup [(A'' \cup A''')' \cup (A'' \cup A'')'],$ by 6 and
$A' \cup A'' = [(A'' \cup A'')' \cup (A'' \cup A')'] \cup [(A''' \cup A'')' \cup (A''' \cup A')']$
whence 9 follows by 3 and 4.

10. $A = (A' \cup A''')' \cup (A' \cup A'')', \ A'' = (A''' \cup A'')' \cup (A''' \cup A')'$ whence the result follows by 3, 9.

11. Let $A \cup A' = X, \ B \cup B' = Y$ then, by 3, 6, 5, 4, 9,
$$\begin{aligned}
Y &= B' \cup B = B' \cup [(B' \cup B')' \cup (B' \cup B)'] \\
&= B' \cup (B'' \cup Y') \\
&= (B' \cup B'') \cup Y' = (B \cup B') \cup Y = Y \cup Y',
\end{aligned}$$
and similarly $\quad X = X' \cup X.$

But by 6, 3, 4
$$\begin{aligned}
Y \cup Y' &= [(Y' \cup X'')' \cup (Y' \cup X')'] \cup [(Y'' \cup X'')' \cup (Y'' \cup X')'] \\
&= [(X'' \cup Y'')' \cup (X'' \cup Y')'] \cup [(X' \cup Y'')' \cup (X' \cup Y')']
\end{aligned}$$
and so, by 6,
$$Y = X' \cup X = X \cup X' = X.$$

12. 1 is unique by 11, and $1 = A' \cup A$ by 3.

13. 0 is unique by 11.

14. From $A' = B'$ follows, by 2, $A'' = B''$ and hence by 10, $A = B$.

15. By 6, $(A' \cup A')' \cup (A' \cup A)' = A$, whence $A'' \cup (A \cup A')' = A$, and so, since $0 = 1' = (A \cup A')'$, we have $A'' \cup 0 = A$ and therefore $0 \cup A = A$, and $A \cup 0 = A$.

16. $A \cap 1 = (A' \cup 1')' = (A' \cup 0)' = A'' = A.$

17. $A \cap A' = (A' \cup A'')' = (A \cup A')' = 1' = 0.$

18. $A \cap B = (A' \cup B')' = (B' \cup A')' = B \cap A.$

19. $(A \cap B) \cap C = (A' \cup B')' \cap C = [(A' \cup B')'' \cup C']'$
$\qquad = [(A' \cup B') \cup C']' = [A' \cup (B' \cup C')]'$
$\qquad = A \cap (B' \cup C')' = A \cap (B \cap C).$

20. $(A' \cap B')' = (A'' \cup B'')'' = A \cup B.$

21. $A \cap A = (A' \cup A')' = A'' = A.$

22. $A \cup 1 = A \cup (A \cup A') = (A \cup A) \cup A' = A \cup A' = 1.$

23. $A \cap 0 = (A' \cup 1)' = 1' = 0.$

24. $A \cup (A \cap B) = (A \cap B) \cup \{(A \cap B) \cup (A \cap B')\}$
$\qquad = (A \cap B) \cup (A \cap B') = A.$

25. $A \cap (A \cup B) = [A' \cup (A \cup B)']' = [A' \cup (A' \cap B')]' = A'' = A.$

26. Since $(A' \cup B')' \cup (A' \cup B)' = A$, if $A' \cup B = 1$ then
$$(A' \cup B')' = A; \text{ moreover, again by 6,}$$
$$(B' \cup A')' \cup (B' \cup A) = B, \text{ and so, if } B' \cup A = 1,$$
$$(B' \cup A')' = B, \text{ whence } A = B.$$

27. From $A \cap B = 0$ and $A \cup B = 1$ follow $B' \cup A' = 1$, $(A')' \cup B = 1$ whence, by 26, $A' = B$.

28. The union in question, by 8, equals
$$(A \cap B) \cup (A \cap B') \cup (A' \cap B) \cup (A' \cap B') = A \cup A' = 1.$$

29. Consider any two terms in the union, e.g. $A \cap B' \cap C$, $A' \cap B \cap C$; their intersection is
$$(A \cap A') \cap (B' \cap C \cap B \cap C) = 0 \cap (B' \cap C \cap B \cap C) = 0.$$

30. $(A \cap B) \cup (A \cap C)$
$$= [(A \cap B \cap C) \cup (A \cap B \cap C')] \cup [(A \cap B' \cap C) \cup (A \cap B \cap C)],$$
by 8,
$$= (A \cap B \cap C) \cup (A \cap B' \cap C) \cup (A \cap B \cap C'), \text{ by 5.}$$

31. $[A \cap (B \cup C)]' = A' \cup (B \cup C)' = A' \cup (B' \cap C')$; but by 8, $A' = (A' \cap B) \cup (A' \cap B')$
$$= (A' \cap B \cap C) \cup (A' \cap B \cap C') \cup (A' \cap B' \cap C) \cup (A' \cap B' \cap C')$$
and $B' \cap C' = (A \cap B' \cap C') \cup (A' \cap B' \cap C')$, whence the result follows.

32. By 30, 31 and 28.

33. Write P, Q, R for the three terms on the right hand side of equation 30, and S, T, U, V, W for those on the right of equation 31, so that
$$(A \cap B) \cup (A \cap C) = P \cup Q \cup R,$$
$$[A \cap (B \cup C)]' = S \cup T \cup U \cup V \cup W; \quad \text{by 29}$$
$$(P \cap S) \cup (P \cap T) = 0 \cup 0, \quad \text{and so}$$
by 32, 15
$$[P \cap (S \cup T)]' = 1, \quad \text{proving} \quad P \cap (S \cup T) = 0.$$
Similarly $P \cap (S \cup T) \cup (P \cap U) = 0 \cup 0 = 0$, so that
$$P \cap (S \cup T \cup U) = 0$$
and so on up to
$$P \cap (S \cup T \cup U \cup V \cup W) = 0.$$
Writing X for $S \cup T \cup U \cup V \cup W$ we have $X \cap P = 0$, and similarly $X \cap Q = 0$, $X \cap R = 0$ and so, exactly as before,
$$X \cap (P \cup Q \cup R) = 0,$$
that is $(P \cup Q \cup R) \cap (S \cup T \cup U \cup V \cup W) = 0$, as required.

34. By 32, 33, 27
$$[(A \cap B) \cup (A \cap C)]' = [A \cap (B \cup C)]'.$$

35. By 10, 7
$$A \cup (B \cap C) = A'' \cup (B' \cup C')' = [A' \cap (B' \cup C')]'$$
$$= [(A' \cap B') \cup (A' \cap C')]'$$
$$= (A' \cap B')' \cap (A' \cap C')'$$
$$= (A \cup B) \cap (A \cup C).$$

This completes the proof in the system based on axioms 1–7 of all the axioms 2.01–2.04; since axioms 1–7 have all previously been proved in axiom system 2.01–2.04 it follows that *these systems are fully equivalent*.

36. If $A \cup B = B$, then
$$A \cap B = (A' \cup B')' = [A' \cup (A \cup B)']' = [A' \cup (A' \cap B')]' = A'' = A;$$
and if
$$A \cap B = A,$$
$$A \cup B = (A' \cap B')' = [(A \cap B)' \cap B']' = (A \cap B)'' \cup B''$$
$$= (A \cap B) \cup B = B.$$

37. If $A \cup B = B$ then
$$A' \cup B = A' \cup (A \cup B) = (A \cup A') \cup B = 1 \cup B = 1.$$
If $A' \cup B = 1$,
$$A \cup B = [A' \cap B']' = [(A' \cap B') \cup 0]' = [(A' \cap B') \cup (B \cap B')]'$$
$$= [(A' \cup B) \cap B']' = (A' \cup B)' \cup B = 1' \cup B = 0 \cup B = B.$$

38. If $A \cup B = B$ then
$$A \cap B' = (A' \cap B')' = [A' \cup (A \cup B)]' = [(A' \cup A) \cup B]' = (1 \cup B)' = 0;$$
if $A \cap B' = 0$, then
$$A' \cup B = (A \cap B')' = 1,$$
whence, by 37, $A \cup B = B$.

39. $A \cap B' = 0$ is equivalent to $A \cup B = B$, $A \cap B = A$, $A' \cup B = 1$, (by 36, 37, 38).

40. If union and complementation satisfy the table

$\cup$	0	1	2	3	4	5	$'$
0	0	1	2	3	4	5	1
1	1	1	1	1	1	1	0
2	2	1	2	1	1	2	3
3	3	1	1	3	3	1	2
4	4	1	1	4	4	1	5
5	5	1	5	1	1	5	4

all axioms are valid except A3, since $5 \cup 2 = 5$, $2 \cup 5 = 2$.

41. Axiom 4 fails since $(2 \cup 1) \cup 3 = 2 \cup 3 = 1$,
$$2 \cup (1 \cup 3) = 2 \cup 0 = 2,$$
but the others are satisfied.

42. Axiom 5 fails since $2 \cup 2 = 1$.

43. Axiom 6 fails since
$$(3' \cup 5')' \cup (3' \cup 5)' = (2 \cup 4)' \cup (2 \cup 5)' = 1' \cup 1' = 0 \neq 3.$$

44. If $X = A + U$, $Y = B + U$ then $X + Y = A + B$.
Conversely, if $X + Y = A + B$, then $Y + B = X + A$
and so $X = A + (B + Y)$, $Y = B + (B + Y)$
showing that $X = A + U$, $Y = B + U$ for the value $B + Y$ of U.

45.1. The complement of this equation is that of example 44.

.2. The general solution is

.21. $X = (C \cup L) \cap N$, $Y = (C \cup M) \cap N$

where L, M, N are arbitrary; for if $C \cap X = C \cap Y$ then

$$X = X \cup (C \cap X) = X \cup (C \cap Y) = (X \cup C) \cap (X \cup Y)$$
$$Y \qquad\qquad = (Y \cup C) \cap (X \cup Y)$$

show that X, Y satisfy .21 with the values $X, Y, X \cup Y$ for L, M, N. Moreover if X, Y are given by .21 then

$$C \cap X = \{C \cap (C \cup L)\} \cap N = C \cap N = \{C \cap (C \cup M)\} \cap N = C \cap Y.$$

.3. The general solution is $X = A' \cap U$, for $A' \cap U \cap A = 0$,

and if $\qquad\qquad A \cap X = 0$ then $A' \cup (A \cap X) = A' \cup 0 = A'$,

whence $\qquad\qquad A' \cup X = A'$ and so $X = X \cap (A' \cup X) = A' \cap X$,

showing that $\qquad\qquad X = A' \cap U$ for the value X of U.

46. If $X = (U \cup A') \cap B$ and $A \subset B$ then

$$A \cup X = A \cup B = B.$$

Conversely if X is a solution of $A \cup X = B$ then $A \subset A \cup X = B$, and $A' \cap X' = B'$, whence $A \cup X' = A \cup (A' \cap X') = A \cup B'$ and so $A' \cap X = A' \cap B$, so that

$$X = X \cup (X \cap A') = X \cup (A' \cap B) = (X \cup A') \cap (X \cup B)$$
$$= (X \cup A') \cap B, \quad \text{since } X \subset B,$$

showing that $X = (U \cup A') \cap B$ for the value X of U.

47. If $A \cap X = B$, $A' \cup X' = B'$ of which the general solution is

$$X' = (U' \cup A) \cap B'$$

i.e. $\qquad\qquad X = (U \cap A') \cup B,$

subject to the condition $A' \subset B'$, i.e. $B \subset A$.

Examples IV

1.1.

p	q	r	$p \to (q \,\&\, r)$	$(p \to q) \,\&\, (p \to r)$
0	0	0	1	1
0	0	1	1	1
0	1	0	1	1
0	1	1	1	1
1	0	0	0	0
1	0	1	0	0
1	1	0	0	0
1	1	1	1	1

with similar solutions to the remaining parts of question 1.

2.4.

p	q	$p \rightarrow q$	$p \rightarrow \neg q$	$(p \rightarrow q) \rightarrow \{(p \rightarrow \neg q) \rightarrow \neg p\}$
0	0	1	1	1
0	1	1	1	1
1	0	0	1	1
1	1	1	0	1

3. From hypotheses $p \rightarrow r$, $q \rightarrow r$ we derive in turn

$$\neg p \lor r, \quad \neg q \lor r, \quad (\neg p \,\&\, \neg q) \lor r, \quad (p \lor q) \rightarrow r$$

whence, by the deduction theorem

$$(p \rightarrow r) \rightarrow \{(q \rightarrow r) \rightarrow [(p \lor q) \rightarrow r]\}.$$

For 2.2 we observe that

$$(p \rightarrow \neg p) \rightarrow \neg p \leftrightarrow p \rightarrow \neg(p \rightarrow \neg p) \leftrightarrow \neg p \lor (p \,\&\, p) \leftrightarrow \neg p \lor p$$

and $p \lor \neg p$ is provable.
Substituting $\neg p$ for p gives 2.3 from 2.2.

$$(p \rightarrow q) \rightarrow \{(p \rightarrow \neg q) \rightarrow \neg p\} \leftrightarrow [(p \rightarrow q) \,\&\, p] \rightarrow \neg(p \rightarrow \neg q)$$
$$\leftrightarrow \neg p \lor (p \,\&\, \neg q) \lor (p \,\&\, q)$$
$$\leftrightarrow (\neg p \lor \neg q) \lor (p \,\&\, q) \leftrightarrow \neg(p \,\&\, q) \lor (p \,\&\, q)$$

which is provable.

4.1. is equivalent to
$$(p \,\&\, \neg q) \lor (q \,\&\, \neg p) \leftrightarrow (p \lor q) \,\&\, (p \lor \neg p) \,\&\, (\neg q \lor q) \,\&\, (\neg p \lor \neg q).$$

4.2. is equivalent to
$$(p \lor \neg q \lor \neg r) \,\&\, (p \lor r \lor \neg r) \,\&\, (\neg q \lor \neg r \lor r)$$
$$\&\, (p \lor \neg q \lor r) \,\&\, (p \lor r) \,\&\, (\neg q \lor r) \,\&\, (\neg r \lor r).$$

4.3. is equivalent to $p \lor q \lor \neg r \lor q$ which is both the disjunctive and conjunctive (one term) normal form.

5. These 15 sentences constitute another axiom system for sentence logic, one in which $\&$, $\lor$, $\neg$, $\rightarrow$ and $\leftrightarrow$ are all signs of the system (and not introduced by definitions).

6.1. follows from the valid sentence $p \rightarrow (q \rightarrow (p \,\&\, q))$
.2. follows from the axiom $p \rightarrow (p \lor q)$
.3. uses example 2.4 and .2 above to derive $P \rightarrow \neg Q$ from $\neg Q$.
.4. uses the equivalence $(p \,\&\, q) \rightarrow r \leftrightarrow p \rightarrow (q \rightarrow r)$.

7. Every sentence is constructed from variables by means of the binary operation $p \lor q$ and negation $\neg p$.
Now $p|p$ has the same truth table as $\neg p$ and $(p|p)|(q|q)$ has the same truth table as $p \lor q$, so that every sentence has a truth table equivalent expressed by means of the function $p|q$ alone. Sentence logic may also be set up as a single axiom system based on this single operation.

8. Writing equality for truth table equivalence we have

$$\neg p = (p, p, p)$$
$$0 = (p, \neg p, r) = (p, (p, p, p), r)$$
$$1 = \neg 0$$
$$p \lor q = \neg(p, q, 1).$$

Examples V

1.1. $a \cap a = a \cap (a \cup (a \cap b)) = a$, by the contraction law.
   $a \cup a = a \cup (a \cap (a \cup b)) = a$.

.2. If
   $a = a \cap b$ then $b = b \cup (a \cap b) = b \cup a$, and conversely.

.3. $\qquad a = a \cup (a \cap b) = a \cup (a \cup b) = (a \cup a) \cup b = a \cup b$
   and $\quad b = b \cup (a \cap b) = b \cup (a \cup b) = (b \cup b) \cup a = a \cup b$.

.4. $a \subset a \cup b \cup c = a \cap b \cap c \subset b$, so that $a \subset b$, and similarly $b \subset a$,
   so that $a = b$, etc.

.5. If $c = c \cap a$ and $c = c \cap b$ then
   $$c \cap (a \cap b) = (c \cap a) \cap (a \cap b) = c \cap a \cap b = (c \cap b) \cap a$$
   $$= c \cap a = c$$
   so that $\qquad\qquad\qquad c \subset a \cap b$.

.6. dual of 1.5.

.7. $a \subset b \to a = a \cap b \to a \cap c = a \cap b \cap c = (a \cap c) \cap (b \cap c)$
   $$\to a \cap c \subset b \cap c.$$

.8. dual of 1.7.

.9. $a \subset b \,\&\, c \subset d \to a = a \cap b \,\&\, c = c \cap d \to a \cap c = (a \cap b) \cap (c \cap d)$
   $$= (a \cap c) \cap (b \cap d)$$
   $$\to a \cap c \subset b \cap d.$$

.91. dual of 1.9.

.92. Since $a \cap b \subset a$ and $a \cap c \subset a$ therefore by 1.6,
   $$(a \cap b) \cup (a \cap c) \subset a;$$
   but $\quad a \cap b \subset b \cup c$ and $a \cap c \subset b \cup c$
   and so $\qquad\qquad (a \cap b) \cup (a \cap c) \subset b \cup c$
   whence 1.92 follows from 1.5.

.93. $a \subset a \cup b$, $a \subset a \cup c$ and so $a \subset [(a \cup b) \cap (a \cup c)]$ whence the
   result follows.

2. Suppose that
   $$a \cap (b \cup c) = (a \cap b) \cup (a \cap c)$$
   then $\quad (a \cup b) \cap (a \cup c) = \{(a \cup b) \cap a\} \cup \{(a \cup b) \cap c\}$
   $$= a \cup \{c \cap (a \cup b)\}$$
   $$= a \cup \{(a \cap c) \cup (b \cap c)\}$$
   $$= \{a \cup (a \cap c)\} \cup (b \cap c)$$
   $$= a \cup (b \cap c)$$
   which is the other distributive law.

3. If $\qquad a = a \cup b \quad$ then $\quad \bar{a} = \overline{a \cup b} = \bar{a} \cup \bar{b}$

and so if $b \subset a$ then $\bar{b} \subset \bar{a}$ which proves that a union homomorphism is also a lattice homomorphism.

4. $a \cap b \to (a \cap k) \cap (b \cap k) = (a \cap b) \cap k$

so that the mapping is an intersection (and so a lattice) homomorphism. Similarly under $a \to a \cup k$,

$$a \cup b \to (a \cup k) \cup (b \cup k) = (a \cup b) \cup k.$$

5. Let $\bar{a}$ be the map of a. Consider the set A of elements of the lattice for which $a \subset \bar{a}$; since the lattice is complete this set has least upper bound, u say. We show that u is unchanged by the mapping, that is $u = \bar{u}$.

Since $a \subset u$ for every $a \, \varepsilon \, A$, and since a homomorphism preserves inclusion, we have $\bar{a} \subset \bar{u}$; but $a \subset \bar{a}$ and so $a \subset \bar{u}$, which shows that $\bar{u}$ is an upper bound of the set A. Since $u \subset \bar{u}$, therefore $\bar{u} \subset \bar{\bar{u}}$, so that, by definition of A, $\bar{u} \, \varepsilon \, A$, and so $\bar{u} \subset u$; but u is the least upper bound, and $\bar{u}$ an upper bound, of A so that $u \subset \bar{u}$, which proves that $u = \bar{u}$.

6. $a \cap b \subset a \cup b, a \cap b \subset b \cup c, a \cap b \subset c \cup a$ so that by 1.5,

$$a \cap b \subset (a \cup b) \cap (b \cup c) \cap (c \cup a) = D, \quad \text{say.}$$

Similarly both $b \cap c, c \cap a$ are contained in D whence the result follows from 1.6.

7. If the lattice is distributive

$\{(a \cup b) \cap (b \cup c)\} \cap (c \cup a)$
$\qquad = \{(a \cup b) \cap (b \cup c) \cap c\} \cup \{(a \cup b) \cap (b \cup c) \cap a\}$
$\qquad = \{(a \cup b) \cap c\} \cup \{a \cap (b \cup c)\}$
$\qquad = \{(a \cap c) \cup (b \cap c)\} \cup \{(a \cap b) \cup (a \cap c)\}$
$\qquad = (a \cap c) \cup (b \cap c) \cup (a \cap b), \quad \text{which proves 7.1.}$

Conversely, if 7.1 holds, writing $(A \cup B) \cap (A \cup C)$ for a, $B \cup C$ for b and A for c in 7.1 the left hand side of 7.1 becomes

$\{(A \cup B) \cap (A \cup C) \cap (B \cup C)\} \cup \{(B \cup C) \cap A\} \cup \{A \cap (A \cup B) \cap (A \cup C)\}$
$\qquad = \{(A \cup B) \cap (A \cup C) \cap (B \cup C)\} \cup A, \quad \text{by the contraction laws,}$
$\qquad = (A \cap C) \cup (B \cap C) \cup (A \cap B) \cup A, \quad \text{by 7.1,}$
$\qquad = A \cup (B \cap C),$

whereas under this substitution the right hand side of 7.1 becomes

$\{[(A \cup B) \cap (A \cup C)] \cup (B \cup C)\} \cap [(B \cup C) \cup A] \cap \{A \cup [(A \cup B) \cap (A \cup C)]\}$
$\qquad = \{[(A \cup B) \cap (A \cup C)] \cup (B \cup C)\} \cap [A \cup B \cup C] \cap [(A \cup B) \cap (A \cup C)],$
$\qquad\qquad\qquad\qquad\qquad\qquad\qquad\qquad\qquad\qquad\qquad \text{by 1.93,}$
$\qquad = \{[(A \cup B) \cap (A \cup C)] \cup (B \cup C)\} \cap [(A \cup B) \cap (A \cup C)],$

since $(A \cup B) \subset (A \cup B \cup C),$

$\qquad = (A \cup B) \cap (A \cup C), \quad \text{by contraction.}$

8. $x = x \cup (x \cap a) = x \cup (a \cap y) = (x \cup a) \cap (x \cup y)$
$\qquad = (a \cup y) \cap (x \cup y) = (a \cap x) \cup y = (a \cap y) \cup y = y.$

9. If $a \subset k$ and $b \subset k$ then $a \cup b \subset k$, (by 1.6), and if $a \subset k$ and $b \subset a$ then $b \subset k$, so that the set of elements contained in k forms an ideal. Let L be a finite lattice in which J is an ideal. J has a finite number of elements $a_1, a_2, \ldots, a_k$ say and $u = a_1 \cup a_2 \cup \ldots \cup a_k$ belongs to J. Thus every element of J is contained in the element u; conversely every element of L contained in u belongs to J, so that J coincides with (u).

10. If we write $m \subset n$ for $m = m \cap n$ then $m \subset n$ says that m divides n and therefore $n = m \cup n$. With respect to this inclusion relation $m \cap n$ is the greatest lower bound of m, n (for any factor of m and n is a factor of their highest common factor), and $m \cup n$ is least upper bound of m, n (for any n divisible by both m and n is divisible by their lowest common multiple).

11. $x \, \varepsilon \, (a) \cap (b) \leftrightarrow x \, \varepsilon \, (a) \, \& \, x \, \varepsilon \, (b) \leftrightarrow x \subset a \, \& \, x \subset b \leftrightarrow x \subset a \cap b$
$\leftrightarrow x \, \varepsilon \, (a \cap b)$.
$x \, \varepsilon \, (a) \cup (b) \leftrightarrow$ there are elements a_1, b_1 of $(a), (b)$ such that $x \subset a_1 \cup b_1$
$\leftrightarrow$ there are elements $a_1 \subset a, b_1 \subset b$ such that $x \subset a_1 \cup b_1$
$\leftrightarrow x \subset a \cup b$
$\leftrightarrow x \, \varepsilon \, (a \cup b)$.

12. If $(a) = (b)$ then, since $a \subset a$, $a \, \varepsilon \, (a)$, $a \, \varepsilon \, (b)$ and so $a \subset b$, and similarly $b \subset a$, proving $a = b$. Thus the mapping of a upon (a) is one-to-one, and by example 11, it preserves the lattice operations and is therefore an isomorphism.

13. $(a + a) = (a + a)(a + a) = aa + aa + aa + aa$, by the distributive law
$$= (a + a) + (a + a)$$
whence
$$0 = (a + a) + \overline{(a + a)} = (a + a) + \{(a + a) + \overline{(a + a)}\}$$
$$= (a + a) + 0 = a + a.$$
It follows that
$$a = a + 0 = a + (a + \bar{a}) = (a + a) + \bar{a} = \bar{a} + 0 = \bar{a}.$$
If a, b are any two elements of a Boolean ring
$$a + b = (a + b)(a + b) = aa + ab + ba + bb = a + ab + ba + b$$
whence $0 = (a + b) + \overline{(a + b)} = ab + ba$, and therefore $ab = \overline{ba}$; but $\overline{ba} = ba$, and so $ab = ba$.

14. The commutative law for $\cap$ and $\cup$ follows from the commutative laws for $a + b$ and ab.
For the associative laws we have only to consider $\cup$:
$$(a \cup b) \cup c = (a + b + ab) + c + (a + b + ab)c$$
$$= a + (b + c + bc) + a(b + c + bc)$$
$$= a \cup (b \cup c).$$
For the contraction laws we have
$$a \cap (a \cup b) = a(a + b + ab) = aa + ab + aab = a + ab + ab = a,$$
$$a \cup (a \cap b) = a + ab + aab = a.$$

For the distributive law we have

$$(a \cap b) \cup (a \cap c) = ab + ac + abac$$
$$= ab + ac + abc$$
$$= a(b + c + bc)$$
$$= a \cap (b \cup c).$$

Finally we note

$$x \cap 0 = x.0 = 0, \qquad x \cup 1 = 1 + x + 1.x = 1,$$
$$x \cap x' = x(x + 1) = xx + x = x + x = 0$$
$$x \cup x' = x + (x + 1) + x(x + 1) = x + x + 1 + xx + x = 1.$$

15. All the necessary relations, namely,

$$a + b = b + a, \qquad a + (b + c) = (a + b) + c$$
$$a \cap (b + c) = (a \cap b) + (a \cap c)$$
$$a + 0 = a, \qquad a + a' = 1$$
$$a + a = a$$

were proved in § 2.67.

16. In the first lattice

$$a \cup (b \cap c) = a \cup 0 = a$$
$$(a \cup b) \cap (a \cup c) = b \cap 1 = b.$$

In the second

$$a \cup (b \cap c) = a \cup c = b$$
$$(a \cup b) \cap (a \cup c) = b \cap b = b, \quad \text{etc.}$$

Apart from a chain of five elements the second and third lattice are the only distributive lattices of five elements.

Bibliography

BIRKHOFF, G., 1948. *Lattice Theory*. 2nd ed. New York. American Mathematical Society Colloquium Publications, Vol. XXV.

DILWORTH, R. P. (Editor), 1948. *Lattice Theory*. New York. American Mathematical Society. Proceedings of Symposia in Pure Mathematics, Vol. II.

HERMES, H., 1955. *Einführung in die Verbandstheorie*. Berlin. Springer-Verlag.

HUNTINGTON, E. V., 1904. Sets of independent postulates for the algebra of logic. *Trans. Amer. Math. Soc.* 5, 208–309.

HUNTINGTON, E. V., 1933. Postulate for the algebra of logic. *Trans. Amer. Math. Soc.* 35, 274–304.

NEWMAN, M. H. A., 1941. A characterization of Boolean lattices and rings. *J. London Math. Soc.* 16, 256–72.

RUDEANU, S., 1959. Boolean equations and their applications to the study of bridge-circuits. 1. *Bull. Math.* 3, 51, No. 4, Bucharest.

STOLL, R. R. 1961. *Sets, Logic and Axiomatic Theories*. San Francisco and London. W. H. Freeman.

STONE, M. H., 1935. Subsumption of Boolean algebras under the theory of rings. *Proc. Nat. Acad. Sci.* 21, 103–5.

STONE, M. H., 1936. The theory of representations for Boolean algebras. *Trans. Amer. Math. Soc.* 40, 37–111.

TARSKI, A., 1935. Zur Grundlagung der Booleschen Algebra. *Fundamenta Mathematica* 24, 177–198.

Index

Absorption laws 24
Associative law 7
Atoms 95
Axioms 21, 49, 74, 78, 92, 115

BOOLE, G. 36
Bound
  lower 97
  upper 97

Chain 95
Class 1
  empty 3
  unit 1
  universal 3
Commutative 4, 21
Comparable elements 94
Complement 3
Complete axiom set 79, 89
Congruence 64
Correspondence 42
Cross 15

Deduction theorem 91
DE MORGAN 6, 27
Denumerable algebra 114
Detachment 79
Difference 11, 29
  symmetric 13, 29
Distributive law 8, 21
Duality 24

Equality 35
Equations, Boolean 55
Equivalence relation 64
Expression, Boolean 32

Finite algebra 109

Homomorphism 43
  lattice 44
HUNTINGTON, E. V. 36

Ideal
  intersection 105
  maximal 111
  union 104
Inclusion 2, 27
Independence of axioms 37, 69, 90
Inference rules 78
Intersection 4
Isomorphism 43

Lattice 98
  complemented 104
  distributive 100
  of ideals 107

Maximal element 95
Membership 1
Minimal element 95

Newman algebra   115
Normal form
  conjunctive   87
  disjunctive   88
Null class   1

Order
  linear   94
  partial   94

Pairs, algebra of   41

Reflexive relation   94
Representation theorems   109, 110

Sentence logic   76
Solution, general   56
Standard forms   30
STONE, M. H.   110

Subclass   2
Substitution   37, 78
Superclass   2
Symmetric difference   13, 29

Test for provability   35
Transitive   50
Truth tables   77

Union   3
Unit   21
  class   1
Universal class   3

Variables   37
Venn diagram   17

Zero   21
Zorn's lemma   111

A CATALOG OF SELECTED
DOVER BOOKS
IN SCIENCE AND MATHEMATICS

Astronomy

BURNHAM'S CELESTIAL HANDBOOK, Robert Burnham, Jr. Thorough guide to the stars beyond our solar system. Exhaustive treatment. Alphabetical by constellation: Andromeda to Cetus in Vol. 1; Chamaeleon to Orion in Vol. 2; and Pavo to Vulpecula in Vol. 3. Hundreds of illustrations. Index in Vol. 3. 2,000pp. 6⅛ x 9¼.
Vol. I: 0-486-23567-X
Vol. II: 0-486-23568-8
Vol. III: 0-486-23673-0

EXPLORING THE MOON THROUGH BINOCULARS AND SMALL TELESCOPES, Ernest H. Cherrington, Jr. Informative, profusely illustrated guide to locating and identifying craters, rills, seas, mountains, other lunar features. Newly revised and updated with special section of new photos. Over 100 photos and diagrams. 240pp. 8¼ x 11. 0-486-24491-1

THE EXTRATERRESTRIAL LIFE DEBATE, 1750–1900, Michael J. Crowe. First detailed, scholarly study in English of the many ideas that developed from 1750 to 1900 regarding the existence of intelligent extraterrestrial life. Examines ideas of Kant, Herschel, Voltaire, Percival Lowell, many other scientists and thinkers. 16 illustrations. 704pp. 5⅜ x 8½. 0-486-40675-X

THEORIES OF THE WORLD FROM ANTIQUITY TO THE COPERNICAN REVOLUTION, Michael J. Crowe. Newly revised edition of an accessible, enlightening book recreates the change from an earth-centered to a sun-centered conception of the solar system. 242pp. 5⅜ x 8½. 0-486-41444-2

A HISTORY OF ASTRONOMY, A. Pannekoek. Well-balanced, carefully reasoned study covers such topics as Ptolemaic theory, work of Copernicus, Kepler, Newton, Eddington's work on stars, much more. Illustrated. References. 521pp. 5⅜ x 8½.
0-486-65994-1

A COMPLETE MANUAL OF AMATEUR ASTRONOMY: TOOLS AND TECHNIQUES FOR ASTRONOMICAL OBSERVATIONS, P. Clay Sherrod with Thomas L. Koed. Concise, highly readable book discusses: selecting, setting up and maintaining a telescope; amateur studies of the sun; lunar topography and occultations; observations of Mars, Jupiter, Saturn, the minor planets and the stars; an introduction to photoelectric photometry; more. 1981 ed. 124 figures. 25 halftones. 37 tables. 335pp. 6½ x 9¼. 0-486-40675-X

AMATEUR ASTRONOMER'S HANDBOOK, J. B. Sidgwick. Timeless, comprehensive coverage of telescopes, mirrors, lenses, mountings, telescope drives, micrometers, spectroscopes, more. 189 illustrations. 576pp. 5⅝ x 8¼. (Available in U.S. only.)
0-486-24034-7

STARS AND RELATIVITY, Ya. B. Zel'dovich and I. D. Novikov. Vol. 1 of *Relativistic Astrophysics* by famed Russian scientists. General relativity, properties of matter under astrophysical conditions, stars, and stellar systems. Deep physical insights, clear presentation. 1971 edition. References. 544pp. 5⅝ x 8¼. 0-486-69424-0

Chemistry

THE SCEPTICAL CHYMIST: THE CLASSIC 1661 TEXT, Robert Boyle. Boyle defines the term "element," asserting that all natural phenomena can be explained by the motion and organization of primary particles. 1911 ed. viii+232pp. 5⅜ x 8½.
0-486-42825-7

RADIOACTIVE SUBSTANCES, Marie Curie. Here is the celebrated scientist's doctoral thesis, the prelude to her receipt of the 1903 Nobel Prize. Curie discusses establishing atomic character of radioactivity found in compounds of uranium and thorium; extraction from pitchblende of polonium and radium; isolation of pure radium chloride; determination of atomic weight of radium; plus electric, photographic, luminous, heat, color effects of radioactivity. ii+94pp. 5⅜ x 8½.   0-486-42550-9

CHEMICAL MAGIC, Leonard A. Ford. Second Edition, Revised by E. Winston Grundmeier. Over 100 unusual stunts demonstrating cold fire, dust explosions, much more. Text explains scientific principles and stresses safety precautions. 128pp. 5⅜ x 8½.   0-486-67628-5

THE DEVELOPMENT OF MODERN CHEMISTRY, Aaron J. Ihde. Authoritative history of chemistry from ancient Greek theory to 20th-century innovation. Covers major chemists and their discoveries. 209 illustrations. 14 tables. Bibliographies. Indices. Appendices. 851pp. 5⅜ x 8½.   0-486-64235-6

CATALYSIS IN CHEMISTRY AND ENZYMOLOGY, William P. Jencks. Exceptionally clear coverage of mechanisms for catalysis, forces in aqueous solution, carbonyl- and acyl-group reactions, practical kinetics, more. 864pp. 5⅜ x 8½.
0-486-65460-5

ELEMENTS OF CHEMISTRY, Antoine Lavoisier. Monumental classic by founder of modern chemistry in remarkable reprint of rare 1790 Kerr translation. A must for every student of chemistry or the history of science. 539pp. 5⅜ x 8½.   0-486-64624-6

THE HISTORICAL BACKGROUND OF CHEMISTRY, Henry M. Leicester. Evolution of ideas, not individual biography. Concentrates on formulation of a coherent set of chemical laws. 260pp. 5⅜ x 8½.   0-486-61053-5

A SHORT HISTORY OF CHEMISTRY, J. R. Partington. Classic exposition explores origins of chemistry, alchemy, early medical chemistry, nature of atmosphere, theory of valency, laws and structure of atomic theory, much more. 428pp. 5⅜ x 8½. (Available in U.S. only.)   0-486-65977-1

GENERAL CHEMISTRY, Linus Pauling. Revised 3rd edition of classic first-year text by Nobel laureate. Atomic and molecular structure, quantum mechanics, statistical mechanics, thermodynamics correlated with descriptive chemistry. Problems. 992pp. 5⅜ x 8½.   0-486-65622-5

FROM ALCHEMY TO CHEMISTRY, John Read. Broad, humanistic treatment focuses on great figures of chemistry and ideas that revolutionized the science. 50 illustrations. 240pp. 5⅜ x 8½.   0-486-28690-8

Engineering

DE RE METALLICA, Georgius Agricola. The famous Hoover translation of greatest treatise on technological chemistry, engineering, geology, mining of early modern times (1556). All 289 original woodcuts. 638pp. 6¾ x 11.   0-486-60006-8

FUNDAMENTALS OF ASTRODYNAMICS, Roger Bate et al. Modern approach developed by U.S. Air Force Academy. Designed as a first course. Problems, exercises. Numerous illustrations. 455pp. 5⅜ x 8½.   0-486-60061-0

DYNAMICS OF FLUIDS IN POROUS MEDIA, Jacob Bear. For advanced students of ground water hydrology, soil mechanics and physics, drainage and irrigation engineering and more. 335 illustrations. Exercises, with answers. 784pp. 6⅛ x 9¼.
0-486-65675-6

THEORY OF VISCOELASTICITY (Second Edition), Richard M. Christensen. Complete consistent description of the linear theory of the viscoelastic behavior of materials. Problem-solving techniques discussed. 1982 edition. 29 figures. xiv+364pp. 6⅛ x 9¼.   0-486-42880-X

MECHANICS, J. P. Den Hartog. A classic introductory text or refresher. Hundreds of applications and design problems illuminate fundamentals of trusses, loaded beams and cables, etc. 334 answered problems. 462pp. 5⅜ x 8½.   0-486-60754-2

MECHANICAL VIBRATIONS, J. P. Den Hartog. Classic textbook offers lucid explanations and illustrative models, applying theories of vibrations to a variety of practical industrial engineering problems. Numerous figures. 233 problems, solutions. Appendix. Index. Preface. 436pp. 5⅜ x 8½.   0-486-64785-4

STRENGTH OF MATERIALS, J. P. Den Hartog. Full, clear treatment of basic material (tension, torsion, bending, etc.) plus advanced material on engineering methods, applications. 350 answered problems. 323pp. 5⅜ x 8½.   0-486-60755-0

A HISTORY OF MECHANICS, René Dugas. Monumental study of mechanical principles from antiquity to quantum mechanics. Contributions of ancient Greeks, Galileo, Leonardo, Kepler, Lagrange, many others. 671pp. 5⅜ x 8½. 0-486-65632-2

STABILITY THEORY AND ITS APPLICATIONS TO STRUCTURAL MECHANICS, Clive L. Dym. Self-contained text focuses on Koiter postbuckling analyses, with mathematical notions of stability of motion. Basing minimum energy principles for static stability upon dynamic concepts of stability of motion, it develops asymptotic buckling and postbuckling analyses from potential energy considerations, with applications to columns, plates, and arches. 1974 ed. 208pp. 5⅜ x 8½.
0-486-42541-X

METAL FATIGUE, N. E. Frost, K. J. Marsh, and L. P. Pook. Definitive, clearly written, and well-illustrated volume addresses all aspects of the subject, from the historical development of understanding metal fatigue to vital concepts of the cyclic stress that causes a crack to grow. Includes 7 appendixes. 544pp. 5⅜ x 8½. 0-486-40927-9

Mathematics

FUNCTIONAL ANALYSIS (Second Corrected Edition), George Bachman and Lawrence Narici. Excellent treatment of subject geared toward students with background in linear algebra, advanced calculus, physics and engineering. Text covers introduction to inner-product spaces, normed, metric spaces, and topological spaces; complete orthonormal sets, the Hahn-Banach Theorem and its consequences, and many other related subjects. 1966 ed. 544pp. 6⅛ x 9¼. 0-486-40251-7

ASYMPTOTIC EXPANSIONS OF INTEGRALS, Norman Bleistein & Richard A. Handelsman. Best introduction to important field with applications in a variety of scientific disciplines. New preface. Problems. Diagrams. Tables. Bibliography. Index. 448pp. 5⅜ x 8½. 0-486-65082-0

VECTOR AND TENSOR ANALYSIS WITH APPLICATIONS, A. I. Borisenko and I. E. Tarapov. Concise introduction. Worked-out problems, solutions, exercises. 257pp. 5⅝ x 8¼. 0-486-63833-2

AN INTRODUCTION TO ORDINARY DIFFERENTIAL EQUATIONS, Earl A. Coddington. A thorough and systematic first course in elementary differential equations for undergraduates in mathematics and science, with many exercises and problems (with answers). Index. 304pp. 5⅜ x 8½. 0-486-65942-9

FOURIER SERIES AND ORTHOGONAL FUNCTIONS, Harry F. Davis. An incisive text combining theory and practical example to introduce Fourier series, orthogonal functions and applications of the Fourier method to boundary-value problems. 570 exercises. Answers and notes. 416pp. 5⅜ x 8½. 0-486-65973-9

COMPUTABILITY AND UNSOLVABILITY, Martin Davis. Classic graduate-level introduction to theory of computability, usually referred to as theory of recurrent functions. New preface and appendix. 288pp. 5⅜ x 8½. 0-486-61471-9

ASYMPTOTIC METHODS IN ANALYSIS, N. G. de Bruijn. An inexpensive, comprehensive guide to asymptotic methods–the pioneering work that teaches by explaining worked examples in detail. Index. 224pp. 5⅜ x 8½ 0-486-64221-6

APPLIED COMPLEX VARIABLES, John W. Dettman. Step-by-step coverage of fundamentals of analytic function theory–plus lucid exposition of five important applications: Potential Theory; Ordinary Differential Equations; Fourier Transforms; Laplace Transforms; Asymptotic Expansions. 66 figures. Exercises at chapter ends. 512pp. 5⅜ x 8½. 0-486-64670-X

INTRODUCTION TO LINEAR ALGEBRA AND DIFFERENTIAL EQUATIONS, John W. Dettman. Excellent text covers complex numbers, determinants, orthonormal bases, Laplace transforms, much more. Exercises with solutions. Undergraduate level. 416pp. 5⅜ x 8½. 0-486-65191-6

RIEMANN'S ZETA FUNCTION, H. M. Edwards. Superb, high-level study of landmark 1859 publication entitled "On the Number of Primes Less Than a Given Magnitude" traces developments in mathematical theory that it inspired. xiv+315pp. 5⅜ x 8½. 0-486-41740-9

CALCULUS OF VARIATIONS WITH APPLICATIONS, George M. Ewing. Applications-oriented introduction to variational theory develops insight and promotes understanding of specialized books, research papers. Suitable for advanced undergraduate/graduate students as primary, supplementary text. 352pp. 5⅜ x 8½.
0-486-64856-7

COMPLEX VARIABLES, Francis J. Flanigan. Unusual approach, delaying complex algebra till harmonic functions have been analyzed from real variable viewpoint. Includes problems with answers. 364pp. 5⅜ x 8½.
0-486-61388-7

AN INTRODUCTION TO THE CALCULUS OF VARIATIONS, Charles Fox. Graduate-level text covers variations of an integral, isoperimetrical problems, least action, special relativity, approximations, more. References. 279pp. 5⅜ x 8½.
0-486-65499-0

COUNTEREXAMPLES IN ANALYSIS, Bernard R. Gelbaum and John M. H. Olmsted. These counterexamples deal mostly with the part of analysis known as "real variables." The first half covers the real number system, and the second half encompasses higher dimensions. 1962 edition. xxiv+198pp. 5⅜ x 8½. 0-486-42875-3

CATASTROPHE THEORY FOR SCIENTISTS AND ENGINEERS, Robert Gilmore. Advanced-level treatment describes mathematics of theory grounded in the work of Poincaré, R. Thom, other mathematicians. Also important applications to problems in mathematics, physics, chemistry and engineering. 1981 edition. References. 28 tables. 397 black-and-white illustrations. xvii + 666pp. 6⅛ x 9¼.
0-486-67539-4

INTRODUCTION TO DIFFERENCE EQUATIONS, Samuel Goldberg. Exceptionally clear exposition of important discipline with applications to sociology, psychology, economics. Many illustrative examples; over 250 problems. 260pp. 5⅜ x 8½.
0-486-65084-7

NUMERICAL METHODS FOR SCIENTISTS AND ENGINEERS, Richard Hamming. Classic text stresses frequency approach in coverage of algorithms, polynomial approximation, Fourier approximation, exponential approximation, other topics. Revised and enlarged 2nd edition. 721pp. 5⅜ x 8½. 0-486-65241-6

INTRODUCTION TO NUMERICAL ANALYSIS (2nd Edition), F. B. Hildebrand. Classic, fundamental treatment covers computation, approximation, interpolation, numerical differentiation and integration, other topics. 150 new problems. 669pp. 5⅜ x 8½. 0-486-65363-3

THREE PEARLS OF NUMBER THEORY, A. Y. Khinchin. Three compelling puzzles require proof of a basic law governing the world of numbers. Challenges concern van der Waerden's theorem, the Landau-Schnirelmann hypothesis and Mann's theorem, and a solution to Waring's problem. Solutions included. 64pp. 5⅜ x 8½.
0-486-40026-3

THE PHILOSOPHY OF MATHEMATICS: AN INTRODUCTORY ESSAY, Stephan Körner. Surveys the views of Plato, Aristotle, Leibniz & Kant concerning propositions and theories of applied and pure mathematics. Introduction. Two appendices. Index. 198pp. 5⅜ x 8½.
0-486-25048-2

INTRODUCTORY REAL ANALYSIS, A.N. Kolmogorov, S. V. Fomin. Translated by Richard A. Silverman. Self-contained, evenly paced introduction to real and functional analysis. Some 350 problems. 403pp. 5⅜ x 8½. 0-486-61226-0

APPLIED ANALYSIS, Cornelius Lanczos. Classic work on analysis and design of finite processes for approximating solution of analytical problems. Algebraic equations, matrices, harmonic analysis, quadrature methods, much more. 559pp. 5⅜ x 8½. 0-486-65656-X

AN INTRODUCTION TO ALGEBRAIC STRUCTURES, Joseph Landin. Superb self-contained text covers "abstract algebra": sets and numbers, theory of groups, theory of rings, much more. Numerous well-chosen examples, exercises. 247pp. 5⅜ x 8½. 0-486-65940-2

QUALITATIVE THEORY OF DIFFERENTIAL EQUATIONS, V. V. Nemytskii and V.V. Stepanov. Classic graduate-level text by two prominent Soviet mathematicians covers classical differential equations as well as topological dynamics and ergodic theory. Bibliographies. 523pp. 5⅜ x 8½. 0-486-65954-2

THEORY OF MATRICES, Sam Perlis. Outstanding text covering rank, nonsingularity and inverses in connection with the development of canonical matrices under the relation of equivalence, and without the intervention of determinants. Includes exercises. 237pp. 5⅜ x 8½. 0-486-66810-X

INTRODUCTION TO ANALYSIS, Maxwell Rosenlicht. Unusually clear, accessible coverage of set theory, real number system, metric spaces, continuous functions, Riemann integration, multiple integrals, more. Wide range of problems. Undergraduate level. Bibliography. 254pp. 5⅜ x 8½. 0-486-65038-3

MODERN NONLINEAR EQUATIONS, Thomas L. Saaty. Emphasizes practical solution of problems; covers seven types of equations. ". . . a welcome contribution to the existing literature...."–*Math Reviews.* 490pp. 5⅜ x 8½. 0-486-64232-1

MATRICES AND LINEAR ALGEBRA, Hans Schneider and George Phillip Barker. Basic textbook covers theory of matrices and its applications to systems of linear equations and related topics such as determinants, eigenvalues and differential equations. Numerous exercises. 432pp. 5⅜ x 8½. 0-486-66014-1

LINEAR ALGEBRA, Georgi E. Shilov. Determinants, linear spaces, matrix algebras, similar topics. For advanced undergraduates, graduates. Silverman translation. 387pp. 5⅜ x 8½. 0-486-63518-X

ELEMENTS OF REAL ANALYSIS, David A. Sprecher. Classic text covers fundamental concepts, real number system, point sets, functions of a real variable, Fourier series, much more. Over 500 exercises. 352pp. 5⅜ x 8½. 0-486-65385-4

SET THEORY AND LOGIC, Robert R. Stoll. Lucid introduction to unified theory of mathematical concepts. Set theory and logic seen as tools for conceptual understanding of real number system. 496pp. 5⅜ x 8¼. 0-486-63829-4

Physics

OPTICAL RESONANCE AND TWO-LEVEL ATOMS, L. Allen and J. H. Eberly. Clear, comprehensive introduction to basic principles behind all quantum optical resonance phenomena. 53 illustrations. Preface. Index. 256pp. 5⅜ x 8½. 0-486-65533-4

QUANTUM THEORY, David Bohm. This advanced undergraduate-level text presents the quantum theory in terms of qualitative and imaginative concepts, followed by specific applications worked out in mathematical detail. Preface. Index. 655pp. 5⅜ x 8½. 0-486-65969-0

ATOMIC PHYSICS (8th EDITION), Max Born. Nobel laureate's lucid treatment of kinetic theory of gases, elementary particles, nuclear atom, wave-corpuscles, atomic structure and spectral lines, much more. Over 40 appendices, bibliography. 495pp. 5⅜ x 8½. 0-486-65984-4

A SOPHISTICATE'S PRIMER OF RELATIVITY, P. W. Bridgman. Geared toward readers already acquainted with special relativity, this book transcends the view of theory as a working tool to answer natural questions: What is a frame of reference? What is a "law of nature"? What is the role of the "observer"? Extensive treatment, written in terms accessible to those without a scientific background. 1983 ed. xlviii+172pp. 5⅜ x 8½. 0-486-42549-5

AN INTRODUCTION TO HAMILTONIAN OPTICS, H. A. Buchdahl. Detailed account of the Hamiltonian treatment of aberration theory in geometrical optics. Many classes of optical systems defined in terms of the symmetries they possess. Problems with detailed solutions. 1970 edition. xv + 360pp. 5⅜ x 8½. 0-486-67597-1

PRIMER OF QUANTUM MECHANICS, Marvin Chester. Introductory text examines the classical quantum bead on a track: its state and representations; operator eigenvalues; harmonic oscillator and bound bead in a symmetric force field; and bead in a spherical shell. Other topics include spin, matrices, and the structure of quantum mechanics; the simplest atom; indistinguishable particles; and stationary-state perturbation theory. 1992 ed. xiv+314pp. 6⅛ x 9¼. 0-486-42878-8

LECTURES ON QUANTUM MECHANICS, Paul A. M. Dirac. Four concise, brilliant lectures on mathematical methods in quantum mechanics from Nobel Prize-winning quantum pioneer build on idea of visualizing quantum theory through the use of classical mechanics. 96pp. 5⅜ x 8½. 0-486-41713-1

THIRTY YEARS THAT SHOOK PHYSICS: THE STORY OF QUANTUM THEORY, George Gamow. Lucid, accessible introduction to influential theory of energy and matter. Careful explanations of Dirac's anti-particles, Bohr's model of the atom, much more. 12 plates. Numerous drawings. 240pp. 5⅜ x 8½. 0-486-24895-X

ELECTRONIC STRUCTURE AND THE PROPERTIES OF SOLIDS: THE PHYSICS OF THE CHEMICAL BOND, Walter A. Harrison. Innovative text offers basic understanding of the electronic structure of covalent and ionic solids, simple metals, transition metals and their compounds. Problems. 1980 edition. 582pp. 6⅛ x 9¼. 0-486-66021-4

CATALOG OF DOVER BOOKS

A TREATISE ON ELECTRICITY AND MAGNETISM, James Clerk Maxwell. Important foundation work of modern physics. Brings to final form Maxwell's theory of electromagnetism and rigorously derives his general equations of field theory. 1,084pp. 5⅜ x 8½. Two-vol. set. Vol. I: 0-486-60636-8 Vol. II: 0-486-60637-6

QUANTUM MECHANICS: PRINCIPLES AND FORMALISM, Roy McWeeny. Graduate student-oriented volume develops subject as fundamental discipline, opening with review of origins of Schrödinger's equations and vector spaces. Focusing on main principles of quantum mechanics and their immediate consequences, it concludes with final generalizations covering alternative "languages" or representations. 1972 ed. 15 figures. xi+155pp. 5⅜ x 8½. 0-486-42829-X

INTRODUCTION TO QUANTUM MECHANICS With Applications to Chemistry, Linus Pauling & E. Bright Wilson, Jr. Classic undergraduate text by Nobel Prize winner applies quantum mechanics to chemical and physical problems. Numerous tables and figures enhance the text. Chapter bibliographies. Appendices. Index. 468pp. 5⅜ x 8½. 0-486-64871-0

METHODS OF THERMODYNAMICS, Howard Reiss. Outstanding text focuses on physical technique of thermodynamics, typical problem areas of understanding, and significance and use of thermodynamic potential. 1965 edition. 238pp. 5⅜ x 8½. 0-486-69445-3

THE ELECTROMAGNETIC FIELD, Albert Shadowitz. Comprehensive undergraduate text covers basics of electric and magnetic fields, builds up to electromagnetic theory. Also related topics, including relativity. Over 900 problems. 768pp. 5⅝ x 8¼. 0-486-65660-8

GREAT EXPERIMENTS IN PHYSICS: FIRSTHAND ACCOUNTS FROM GALILEO TO EINSTEIN, Morris H. Shamos (ed.). 25 crucial discoveries: Newton's laws of motion, Chadwick's study of the neutron, Hertz on electromagnetic waves, more. Original accounts clearly annotated. 370pp. 5⅜ x 8½. 0-486-25346-5

EINSTEIN'S LEGACY, Julian Schwinger. A Nobel Laureate relates fascinating story of Einstein and development of relativity theory in well-illustrated, nontechnical volume. Subjects include meaning of time, paradoxes of space travel, gravity and its effect on light, non-Euclidean geometry and curving of space-time, impact of radio astronomy and space-age discoveries, and more. 189 b/w illustrations. xiv+250pp. 8⅜ x 9¼. 0-486-41974-6

STATISTICAL PHYSICS, Gregory H. Wannier. Classic text combines thermodynamics, statistical mechanics and kinetic theory in one unified presentation of thermal physics. Problems with solutions. Bibliography. 532pp. 5⅜ x 8½. 0-486-65401-X

Paperbound unless otherwise indicated. Available at your book dealer, online at **www.doverpublications.com**, or by writing to Dept. GI, Dover Publications, Inc., 31 East 2nd Street, Mineola, NY 11501. For current price information or for free catalogues (please indicate field of interest), write to Dover Publications or log on to **www.doverpublications.com** and see every Dover book in print. Dover publishes more than 500 books each year on science, elementary and advanced mathematics, biology, music, art, literary history, social sciences, and other areas.